Mushrooms of North America **in Color**

Leucopholiota decorosa

Mushrooms of North America **in Color**

A Field Guide Companion to Seldom-Illustrated Fungi

Alan E. Bessette

Orson K. Miller, Jr.

Arleen R. Bessette

Hope H. Miller

SU Syracuse University Press

First Edition 1995
95 96 97 98 99 00 6 5 4 3 2 1

Photo Credit: All photographs were taken by
Alan and Arleen Bessette and Orson Miller.

The paper used in this publication meets the minimum requirements of American National Standard for Information Sciences—Permanence of Paper for Printed Library Materials, ANSI Z39.48–1984. ∞™

Library of Congress Cataloging-in-Publication Data

Mushrooms of North America in color : a field guide companion to seldom-illustrated fungi / Alan E. Bessette . . . [et al.] — 1st ed.
p. cm.
Includes bibliographical references (p.) and index.
ISBN 0-8156-2666-5 (cloth) : alk. paper). —
ISBN 0-8156-0323-1 (pbk. : alk. paper)
1. Mushrooms — North America — Identification. 2. Mushrooms—North America—Pictorial works. I. Bessette, Alan.
QK617.M84 1995
589.2'22'097—dc2220 95-11781

Book Design and production by Hillside Studio, Inc.
Printed in Hong Kong by Everbest Printing Company, Ltd.

Contents

Illustrations

Preface

DURING THE LAST DECADE, many field guides for mushroom identification have been published. Unfortunately, the majority of the species described and illustrated are the most common species, and many are duplicated in each new book. Few new species are described and accompanied by a color photograph in the popular guides. Publishers have routinely insisted on including those species already commonly illustrated in other field guides, as well as those that are edible, in order to be competitive in the market place and to appeal to a large audience.

Of the more than 5,000 species of mushrooms estimated to occur in North America, only about 2,000 have ever been illustrated using color photographs. Scientific journals have published original descriptions of many genera and species accompanied only by a black and white photograph or drawing at best, and often with no illustration at all. Some descriptions have been published in scientific journals and other sources that are seen by, and available to, only a small number of people. Most major works, especially monographs, often describe many species without the benefit of color illustrations. Still other works, especially field guides, contain many species illustrated in color but with short, nontechnical descriptions.

The objective of this work is to provide for each species an accurate description, a color illustration of high quality, and information on the distinctive characteristics which make it unique. It is meant to be used by amateur and professional mycologists, ecologists, and botanists. The species selected are those which are uncommon, rare or not well illustrated in the current literature. The objectives of the authors are to increase the user's knowledge of these species and to establish a reference source to which individuals can turn for help in identification. Our goal is to provide an extension of the popular field guides already available throughout North America.

Mushrooms of North America in Color presents an accurate description of each illustrated species, including morphological features, spore print, macrochemical tests when applicable, microscopic features, information on habit, habitat, distribution, fruiting data, and frequency. We have also included observations and notes on key identification features, similar species, and sources for further information. The species included in this work are found in a variety of habitats in North America and were selected without emphasizing any particular genus. Descriptions are arranged alphabetically according to orders, with families, genera and species appearing alphabetically within each order presented. Every effort has been made to write the descriptions in clear and easy-to-follow language. It has, however, been necessary to use some technical terms to achieve an acceptable level of accuracy. Each technical term used is defined in the accompanying glossary.

It is our hope that this work will become a companion to the popular field guides. In addition, it will serve as a resource that will assist the user in the identification of additional species that may have previously gone unnamed and unillustrated.

Acknowledgments

WE WOULD LIKE TO THANK Leeds and Marie Bailey, Robert Farnsworth and Beverly Farnsworth for valuable mycological contributions. We are grateful to the late Ellen Trueblood, who discovered a number of rare taxa, some of which are treated here. Thanks also to Dr. Joseph Ammirati, William R. Burk, Dr. Robert Gilbertson, Dr. Gáston Guzmán, Dr. Richard Homola, Dr. Egon Horak, Dr. Gregory Mueller, Dr. Clark Ovrebo, Bill Roody, Walt Sturgeon, Dr. Tom Volk and Ben Woo for mycological notes, technical information, and assistance with species identification. We thank the members of the mushroom clubs who have invited us to share their fungi and their knowledge of them. We greatly appreciate the efforts and contributions of Allein Stanley and Dr. Donald M. Huffman, who reviewed the manuscript and made valuable comments and suggestions for its improvement. We are especially grateful to Dr. Robert Mandel and his staff at the Syracuse University Press who made this book possible.

Although we have provided information regarding the edibility of various mushrooms, this book is not intended to function as a field guide for the accurate identification and safe consumption of wild mushrooms. Readers interested in consuming wild fungi should consult other sources of information, including experienced mycologists or other literary sources or both, before eating any wild mushrooms. Neither the authors nor the publisher are responsible for adverse reactions that may result from eating wild mushrooms.

Species Descriptions and Illustrations

Amanita armillariiformis Trueblood and Jenkins, *Mycologia* 82: 120–123. 1990.

Cap: 1⅝–6⅜″ (4–16 cm) broad, convex to plane at maturity, glabrous, dry but very slightly viscid at first, dull white to dull pinkish buff; universal veil often absent, when present as thin patches or crusts with a pale salmon tint on low polygonal to irregular areolae; margin incurved at first with a distinct sterile edge and scanty remains of the thin, white, partial veil.

Gills: nearly free or joined by a fine line, subdistant, broad, with two tiers of lamellulae, cream color to pinkish and pale tan at maturity.

Stalk: 1⅜–3½″ (3.5–9 cm) long, ⅝–⅞″ (1.7–2.4 cm) wide, equal or expanding toward the base, nearly glabrous and white at apex with fine tufts of fibrils and scales over lower surface; partial veil thin, membranous, white, superior, forming an annular ring, adhering in part to the cap margin or falling away completely at maturity; universal veil remnants on lower stalk remain as white often tinted salmon, concentric rings, partial rings, or patches.

Flesh: white, very firm, homogeneous in cap and stalk; odor unpleasant, often strong and medicinal; taste not attempted.

Technical Features: pileipellis a dense nongelatinized mixocutis of thin-walled hyphae 2.3–8.5 μm diameter; **pileotrama** of loosely interwoven hyphae 2.5–9 μm diameter mixed with inflated cells up to 140 × 37 μm; **lamellar trama** bilateral with a distinct mediostratum, hyphae 2.7–12 μm diameter, **partial veil** mostly filamentous hyphae 2–11 μm diameter with scattered inflated cells; **universal veil** of filamentous hyphae 3.5–11 μm diameter intermixed with inflated, thin-walled, clavate to fusiform cells up to 141 × 34 μm; **clamp connections** present in all tissues; **pseudocystidia** on face and edge of gills elliptical to fusiform, up to 129 × 31μm; **basidia** up to 62 × 4–11 μm, clavate, thin-walled, 4-spored with basal clamp; **spores** 10–13 × 6.2–8.2 μm, elliptic to somewhat elongate, thin-walled, amyloid.

Spore Print: white.

Fruiting: one to several on the ground in dry habitats among sagebrush, mustards, cheat grass, but usually near aspen, Douglas fir, or willows in late March and April to mid-June; distributed from Owyhee County, Idaho west to Ontario, Oregon; infrequent.

Edibility: unknown.

Observations: The robust, white cap up to 6⅜″ (16 cm) wide, patches of universal veil with a pale salmon tint, sterile margin, cream to pink or tan gills, membranous partial veil, distinctive pseudocystidia, amyloid elliptical spores 10–13 × 6.2–8.2 μm and the unique habitat are distinctive characters of *A. armillariiformis*. *Armillaria malheurensis* has a pale tan cap, universal veil with a brownish tint, cream colored gills, and similar amyloid spores and pseudocystidia (Miller et al. 1990). This latter species found in even dryer habitats in a more restricted area in Malheur County, Oregon may in fact be a variant of *A. armillariiformis*. Only additional fresh collections and habitat study can resolve this problem.

Agaricales
Amanitaceae

Amanita pantherina var. **multisquamosa** (Pk.) Jenkins, *Biblio. Mycol.* 57: 65–69. 1977.

Amanita multisquamosa Pk., *Ann. Rep. N.Y. State Mus.* 53: 840. 1900.

Amanita cothurnata Atk., *Stud. Amer. Fungi,* 66–69. 1900.

Cap: 1⅜–4½″ (3.5–11.5 cm) wide, convex to nearly flat and often slightly depressed at maturity; surface white to pale cream with a tan disc, smooth, slightly sticky, with soft, white patches or warts that are easily removed; margin incurved when young, striate to tuberculate-striate.

Gills: deeply adnexed, becoming free at maturity, white, crowded, moderately broad, with numerous truncate lamellulae.

Stalk: 1⅜–5⅛″ (3.5–13 cm) long, ¼–¾″ (7–20 mm) thick, enlarging downward, whitish, covered with cottony fibrils or scales, typically hollow; partial veil white, floccose-membranous; ring superior to median, whitish, floccose-membranous, often flaring then pendant; volva whitish, forming a distinct collar, enclosing an oval to round basal bulb; buttons egg-shaped, enclosed by a whitish, floccose-membranous universal veil.

Flesh: white, thick, firm; odor and taste not distinctive.

Technical Features: pileipellis of interwoven filamentous hyphae, slightly to strongly gelatinized; **lamellar trama** bilateral; **basidia** 37–62 × 4–11 µm, 4-spored; **spores** 8.5–11.8 × 6.3–8.7 µm, round to elliptical, smooth, thin-walled, hyaline, inamyloid.

Spore Print: white.

Fruiting: scattered or in groups on soil in mixed conifer and hardwoods; July–October; Maine to Florida, west to Michigan; common.

Edibility: poisonous.

Observations: The distinctive features of this species include a whitish volva forming a distinct collar at the top of the basal bulb, a distinctly striate margin, and a white to cream cap with a tan disc. Several varieties of *A. pantherina* have been described and are separated primarily by cap color and the appearance of the volva.

Notes: This species is also known as *A. multisquamosa* Pk. and more commonly called *A. cothurnata* Atk. Jenkins (1977) placed these two taxa in synonymy and recognized them as a variety of *A. pantherina*. Therefore, in accordance with the International Code of Botanical Nomenclature, the name *multisquamosa* must be used. For additional information see Jenkins (1986, 40).

Amanita wellsii (Murr.) Sacc., *Syll. Fung.* 13: 2–3. 1925.

Cap: 1⅜–4¾″ (3.5–12 cm) wide, convex to nearly flat, yellowish orange to pinkish orange, sometimes pale orange-brown or yellowish in age; surface dry to slightly sticky when moist, covered with soft, yellowish scale-like patches that are easily removed; margin incurved at first then expanded, nonstriate when young, occasionally slightly striate when mature, with pale yellow lacerated flaps of tissue extending beyond the gills.

Gills: deeply adnexed to free, creamy white, crowded, moderately broad, edges serrate to lacerate, with several tiers of truncate lamellulae.

Stalk: 3–6½″ (7.5–16.5 cm) long, ¼–1″ (7–25 mm) thick, tapering toward the apex, pale yellow, sometimes orange or white at the base, becoming pale orange-brown where handled or in age, fibrillose-scaly at least on the upper half, becoming stuffed to hollow in age; partial veil pale yellow, membranous, typically adhering to the cap margin, sometimes forming a ring; ring superior, pale yellow, membranous, very fragile, evanescent, typically absent because partial veil usually adheres to the cap margin; volval remnants pale yellow, fragile, quickly falling off, enclosing a round to oval basal bulb; buttons egg-shaped, enclosed by a yellowish, cottony, universal veil.

Flesh: pale yellow, thick, firm; odor and taste not distinctive.

Technical Features: **pileipellis** of filamentous interwoven hyphae, slightly gelatinized; **lamellar trama** bilateral; **basidia** 45–65 × 4–11µm, 4-spored; **spores** 11–14 × 6.3–8.3µm, elliptical, smooth, thin-walled, hyaline, inamyloid.

Spore Print: white.

Fruiting: solitary, scattered or in groups on soil in mixed woods; August–September; Quebec to North Carolina; infrequent.

Edibility: unknown.

Observations: The combination of orangish cap colors and a fibrillose-scaly yellowish stalk makes this species easy to identify.

Notes: For additional information see Jenkins (1986, 56).

Boletus pulverulentus Opat., *Wiegmaan's Archiv. Naturgesch* 2: 27. 1836.

Cap: 1½–5″ (4–12.5 cm) wide, pulvinate to convex then broadly convex to nearly plane, dark yellow-brown to blackish brown, becoming dark cinnamon-brown and often developing reddish tints in age; surface dry, subtomentose, usually areolate at maturity; margin entire, often extending slightly over the pore surface.

Pore Surface: bright yellow when young, darkening to golden yellow when mature, instantly staining dark blue when bruised; pores medium to large, up to 1⁄16″ (1.5 mm) in diameter, angular, unequal, elongated near the stalk; tubes adnate at first then depressed around the stalk, ¼–⅝″ (6–16 mm) deep, yellow, instantly staining blue when exposed.

Stalk: 1⅜–3½″ (3.5–9 cm) long, ⅝–1⅜″ (1.5–3.5 cm) thick, enlarging toward the base or nearly equal, solid, pruinose at the apex when young, becoming nearly smooth overall in age, yellow to orange-yellow, typically reddish brown and pruinose toward the base, quickly staining dark blue when handled.

Flesh: yellow in cap, thick, spongy, instantly changing to dark blue upon exposure; stalk reddish brown in base, yellow above and instantly staining dark blue when exposed; odor and taste not distinctive.

Technical Features: addition of ammonium hydroxide to the pileipellis produces a green flash reaction; **pleurocystidia** 30–60 µm, fusoid-ventricose, smooth, thin-walled, with dextrinoid amorphous contents in Melzer's solution; **cheilocystidia** identical to the pleurocystidia; **basidia** mostly 4-spored; **spores** 11–15 × 4–6 µm, fusoid to elliptical, smooth, thin-walled.

Spore Print: dark olive to olive-brown.

Fruiting: scattered or in groups on soil under conifers or hardwoods; July–November; Nova Scotia and Ontario south to Florida, west to Michigan, also the Pacific Northwest; infrequent.

Edibility: edible.

Observations: The dark cap color, yellow stalk with reddish brown near the base, mild-tasting flesh, and quick staining reactions make this species easy to identify.

Notes: For additional information see Smith and Thiers (1971, 312–313).

Boletus roxanae Frost, *Bull. Buffalo Soc. Nat. Sci.* 2: 104. 1874.
Xerocomus roxanae (Frost) Snell, *Mycologia* 37: 383. 1945.

Cap: 1⅛–3½″ (3–9 cm) wide, convex to broadly convex, becoming nearly plane in age, rusty red to dark orange-yellow, fading somewhat in age; surface unpolished, dry, granular-scaly when young, becoming nearly smooth at maturity; margin typically paler than the disc, entire.

Pore Surfaces: whitish, becoming pale yellow and finally pale golden yellow in age, darkening somewhat or developing pale cinnamon stains when injured; pores small to medium, 1–3 per mm, round; tubes adnate at first then depressed around the stalk, ¼–⅜″ (6–10 mm) deep, pale yellow, not staining.

Stalk: 1–3½″ (2.5–9 cm) long, ¼–⅝″ (7–16 mm) thick, enlarging toward the base, solid, faintly pruinose and longitudinally striate, at least on the upper portion, pale orange-yellow, typically with a distinct dull orange zone at the apex.

Flesh: whitish to pale yellow in cap, moderately thick, firm, not staining when exposed; stalk yellowish; a drop of 3% KOH produces a pale brown discoloration; odor not distinctive or weakly pungent; taste unpleasant.

Technical Features: pileipellis a trichodermium of thin-walled hyphae 5–12 µm wide, with some cells 10–28 µm wide; **clamp connections** absent; **pleurocystidia** scattered, 30–46 × 8–12 µm, fusoid-ventricose, thin-walled, smooth; **basidia** 4-spored; **spores** 8–13 × 3–5 µm, ellipsoid to fusiform-ellipsoid, smooth, thin-walled.

Spore Print: dark olive-brown.

Fruiting: singly or scattered on the ground under oaks; August–October; eastern Canada to North Carolina, west to Michigan; occasional.

Edibility: unknown.

Observations: The rusty red to dark orange-yellow cap, a white pore surface which becomes yellow in age, and a pale orange-yellow stalk with a distinct dull orange zone at the apex are the distinctive features. *Boletus roxanae* var. *auricolor* Pk. has a bright yellow stalk and a yellow cap.

Notes: For additional information see Smith and Thiers (1971, 234–236).

Fuscoboletinus paluster (Pk.) Pomerleau, *Mycologia* 56: 708–711. 1964.

Boletus paluster Pk., *Ann. Rep. N.Y. State Mus.* 23: 132. 1872.

Boletinus paluster (Pk.) Pk., *Bull. N.Y. State Mus.* 8: 78. 1889.

Boletinellus paluster (Pk.) Murr., *Mycologia* 1: 8. 1909.

Cap: ¾–2¾″ (2–7 cm) wide, broadly convex to plane or slightly depressed, with or without an umbo; surface dry, finely tomentose to fibrillose-scaly, pale pinkish purple to reddish purple; margin entire, thin, incurved when young, typically supporting remnants of the partial veil on immature specimens.

Pore Surface: pale yellow when young, becoming golden yellow and finally brownish yellow in age, not blueing when injured; pores large, angular, and radially arranged when young, becoming gill-like with crossveins at maturity; tubes strongly decurrent, ¹⁄₁₆–⅛″ (1.5–3 mm) deep, pale yellow then brownish yellow in age, unchanging on exposure.

Stalk: ¾–2″ (2–5 cm) long, ⅛–5⁄16″ (3–7 mm) thick, nearly equal, often crooked, solid; surface finely pruinose to fibrillose, yellow at the apex, pale pinkish purple to reddish purple below, with a yellow basal mycelium; partial veil reddish purple, fibrillose.

Flesh: yellowish white to yellow in cap, reddish under the pileipellis, thin, soft, unchanging when exposed; white in stalk with a yellow unchanging base; odor not distinctive; taste mild to slightly acidic.

Technical Features: pileipellis a trichodermium with end cells 10–16 µm thick; **pleurocystidia** subcylindric, thin-walled, with rounded apices, of two sizes: 80–110 × 8–12 µm or 50–75 × 9–11 µm; **cheilocystidia** cylindric, clavate to subventricose, thin-walled, 20–80 × 6–12 µm; **basidia** 25–45 µm, elongate-clavate to subcylindric, mostly 4-spored; **spores** 7–10 × 3–4 µm, elliptical to subelliptical, smooth, thin-walled.

Spore Print: purple-brown to pinkish brown.

Fruiting: scattered or in groups, usually among sphagnum mosses, under larch; August-November; eastern Canada south to Pennsylvania, west to Wisconsin; occasional to fairly common.

Edibility: edible.

Observations: The purplish cap and stalk and yellowish gill-like pore surface make this one of the most beautiful and easily identified boletes in eastern North America.

Notes: For more information see Pomerleau (1964), and Smith and Thiers (1971, 85–87).

Fuscoboletinus serotinus (Frost) A. H. Sm. and Thiers, *The Boletes of Michigan,* 85–87. 1971.

Boletus serotinus Frost, *Bull. Buffalo Soc. Nat. Sci.* 2: 100. 1874.

Cap: 1½–4¾″ (4–12 cm) wide, convex, becoming broadly convex to plane, often with a low umbo, covered by a dark reddish brown slime layer, fading to pale reddish brown or yellow-brown in age; surface sticky to slimy when fresh, shiny when dry; margin bearing cottony pieces of grayish white, torn partial veil.

Pore Surface: whitish when young, becoming grayish white then pale reddish brown in age, staining purplish gray then reddish brown when injured; pores small to medium, 1–3 per mm, angular, not radially arranged; tubes slightly decurrent, ¼–⅝″ (9–16 mm) deep, whitish to grayish white, staining purplish gray then reddish brown when injured.

Stalk: 2–4″ (5–10 cm) long, ¼–⅝″ (7–16 mm) thick, nearly equal overall, dingy white and weakly reticulate above the ring, whitish with pinkish brown to yellowish brown streaks below; partial veil cottony-membranous, grayish white; ring superior, cottony-membranous, grayish white.

Flesh: in cap white to pale yellow, slowly staining bluish then purplish gray and finally reddish brown, quickly green in $FeSO_4$; in stalk whitish except yellow at base, slowly staining blue when exposed; odor somewhat pungent; taste not distinctive.

Technical Features: pileipellis of hyphae 3–6 µm wide in a gelatinous matrix, hyphae with pale vinaceous contents when mounted in KOH; **pleurocystidia** 55–80 × 6–8 µm, cylindric to subclavate; **cheilocystidia** 32–60 × 5–6 µm, nearly identical to the pleurocystidia; **basidia** 4-spored; **spores** 8–12 × 4–5 µm, oblong to subelliptic, with a shallow suprahilar depression, smooth.

Spore Print: purplish brown.

Fruiting: scattered or in groups on the ground under larch, eastern white pine and other conifers, especially near bogs; August–October; northeastern North America; infrequent.

Edibility: edible.

Observations: The distinctive features include a slimy dark reddish brown cap with cottony, grayish white torn veil pieces on the margin, a whitish pore surface on young specimens and a whitish stalk with a grayish white, cottony-membranous, superior ring. Wrapped specimens usually stain waxed paper blue in less than 30 minutes. *Fuscoboletinus viscidus* (L. ex Fr. et Hoek) Grund and Harrison = *Suillus laricinus* (Berk.). Kuntze is very similar but lacks the dark reddish brown slime layer on the cap and its flesh stains only bluish green, not purplish gray or reddish brown. This mushroom may be confused with *Suillus* species because of its sticky to slimy cap, but differs because it has a purplish brown spore print and lacks glandular dots on its stalk.

Notes: For additional information see Smith and Thiers (1971, 85–87).

Leccinum fibrillosum A. H. Sm., Thiers, and Watling, *Mich. Bot.* 5: 165–166. 1966.

Cap: 2⅜–5½″ (6–25 cm) broad, broadly convex to nearly plane in age, dry to sticky in moist weather, with small squamules or matted fibrils, the fibrils dark reddish brown, darker than the light reddish brown ground color, margin appendiculate with 1–2 mm of sterile tissue.

Pore Surface: buff to light olivaceous when young, becoming olive-brown in age, staining brown when bruised; pores at first 3–4 per mm then 1–2 per mm in age.

Stalk: 2¾–4⅜″ (7–11 cm) long, 1–1¾″ (2.5–4.5 cm) wide, expanding toward the base, with dark blackish brown scabers and irregular reticulation over a white ground color.

Flesh: white, firm, immediately flushing pink in cap and upper stalk when cut and bruised, then changing to muddy bluish red to bluish purple or fuscous in 3 to 4 minutes, often nearly black after several minutes; fresh tissue greenish blue in $FeSO_4$; odor not distinctive; taste pleasant.

Technical Features: **pileipellis** a trichodermium of erect or tangled hyphae 6–12 μm diameter, end cells somewhat cystidioid; **pleurocystidia** and **cheilocystidia** 35–43 × 7–8.5 μm, fusiform to fusoid-ventricose, thin-walled, hyaline to dark yellow-brown in 3% KOH; **caulocystidia** 23–60 × 8–18 μm, thin-walled, fusiform to ovate, often pointed; **clamp connections** absent; **basidia** 23–36 × 8–11 μm, clavate, thin-walled, 4-spored; **spores** 14–18 × 3.8–5 μm, elliptical, thin-walled, yellow-brown in 3% KOH and Melzer's solution.

Spore Print: olive-brown.

Fruiting: several to gregarious under conifers, especially Engelmann spruce and lodgepole pine following thunder showers and rain in late July through early fall; distributed in the Rocky Mountains especially Colorado, northwestern Montana, Idaho, western Wyoming, eastern Washington, and Alberta, Canada; usually infrequent but common after August rains.

Edibility: edible.

Observations: The conspicuously dark reddish brown caps with blackish brown squamules or fibrils, robust appearance, pink to fuscous reaction of the flesh when cut and bruised, and presence under conifers in the Rocky Mountains is a distinctive combination of characters. There is a large number of species of *Leccinum* (Smith and Smith 1973) in North America but very few are associated with conifers. The exceptional year with strong August thunder showers produces large numbers of fruiting bodies.

Phylloporus boletinoides A. H. Sm. and Thiers, *A Contribution Toward a Monograph of North American Species of Suillus,* 105–106. 1964.

Cap: ¾–2″ (2–5 cm) wide, broadly convex when young, becoming nearly plane and sometimes shallowly depressed in age, cinnamon to dark pinkish brown, fading to dull yellow-brown in age; surface dry, velvety tomentose to minutely squamulose, becoming nearly smooth at maturity; margin strongly incurved when young and remaining so at maturity, entire, extending over the pore surface.

Pore Surface: pale olive-buff when young, becoming dark olive-buff at maturity, sometimes staining dark blue to bluish green when bruised but usually not bluing at all; pores strongly boletinoid to sublamellate, ⅛–¼″ (3–6 mm) or sometimes longer; tubes extending well down on the stalk, ⅛–³⁄₁₆″ (3–4.5 mm) deep, pale yellow to olive-buff, dark olive-buff in age.

Stalk: 1–2⅜″ (2.5–6 cm) long, ¼–⅝″ (7–16 mm) thick, tapering slightly downward, solid or hollow in the base, smooth, dry, pale yellow at apex, pale cinnamon below, with a sparse layer of pale yellow basal mycelium.

Flesh: white to whitish in cap, moderately thick, slowly staining grayish when exposed or bruised but not bluing; in stalk pale yellow, not staining; odor not distinctive; taste mild to slightly acidic.

Technical Features: **pileipellis** a trichodermium of upright but tangled hyphae 7–11 µm in diameter; **pleurocystidia** 40–60 × 10–17 µm, thin-walled, smooth, fusoid-ventricose varying to subcylindric or narrowly clavate, apices obtuse; **cheilocystidia** scattered to numerous, similar to the pleurocystidia; **basidia** 4-spored; **spores** 11–16 × 5–6.5 µm, subcylindric to narrowly oval, smooth, thin-walled.

Spore Print: olive-brown.

Fruiting: solitary or scattered on the ground in mixed pine and oak woods; July-December; along the East Coast from Maine to Florida, and west to Texas; infrequent in its northern range but fairly common southward.

Edibility: unknown.

Observations: The distinctive features include a brownish cap with an incurved margin when young, an olive-buff pore surface which sometimes stains blue to green, and the boletinoid to sublamellate pores. The photograph of *P. foliiporus* (Murr.) Phillips and Kibby shown in *Mushrooms of North America* (Phillips 1991, 217) is *P. boletinoides* A. H. Sm. and Thiers.

Notes: For additional information see Smith and Thiers (1964, 105–106).

Suillus albivelatus A. H. Sm., Thiers, and O. K. Miller, *Lloydia* 28: 121–123. 1965.

Cap: 1½–4½″ (4–12 cm) broad, broadly convex, plane in age, viscid to glutinous, glabrous, buttons tinted orange-brown at first, vinaceous-brown or brownish red to yellow-brown in age; margin with scattered white squamules or merely appendiculate with the scanty remains of the white, partial veil.

Pore Surface: buff to orange-buff in age, yellow in cross section; pores emarginate, 1–3 per mm, round, somewhat boletinoid; tubes 3⁄16″ (4.5 mm) long.

Stalk: 3⁄8–1 5⁄8″ (1–4 cm) long, 5⁄8–1 5⁄16″ (1.5–3.3 cm) wide, equal, dry, white, with remains of the white, superior, partial veil which leaves a ragged annular zone which soon disappears, yellowish to warm buff at apex, white below, smooth or with a few glandular dots which develop in age, usually near the apex.

Flesh: firm, white becoming buff to lemon-yellow in age especially near the pores, white to light buff in age, worm holes staining pinkish; odor pleasant; taste mild and of good flavor.

Technical Features: pileipellis of appressed hyphae 4–9 µm diameter, hyaline, in a gelatinous matrix; **clamp connections** present on all tissues; **cystidia** 38–50 × 8–12 µm, subfusoid, cylindric or clavate, thin-walled, vinaceous-red in 3% KOH; **basidia** 26–43 × 6.4–8 µm, clavate, thin-walled, 4-spored; **spores** 7–8.5 × 2.5–3 µm, elliptical, thin-walled, inamyloid.

Spore Print: cinnamon-buff to brown.

Fruiting: single to gregarious under conifers, especially ponderosa pine and other pines; in early spring, occasionally during a wet summer and again in the fall; known only from the Pacific Northwest and the Rocky Mountains in Idaho and northwestern Montana; infrequent to frequent.

Edibility: edible.

Observations: The short and squatty appearance, white partial veil, combined with the vinaceous-red, viscid cap, the white stalk with very few or no glandular dots, and the association with pines in the western United States and Canada are the distinctive characters. *Suillus albivelatus* reminds one of *S. brevipes* (Pk.) Kuntze with a partial veil (see Miller and Miller 1980, Fig. 39). In age, a careful inspection of the margin usually reveals some remains of the white partial veil. On one occasion *S. albivelatus* was collected under open grown pure ponderosa pine in Glacier National Park.

Notes: A color photograph of *S. albivelatus* by Phillips (1991) shows some minute remains of the partial veil, but the veil is not illustrated in young specimens.

Cortinarius mutabilis A. H. Sm., *Lloydia* 7(3): 190–191. 1944.

Cap: 1½–4″ (4–10 cm) wide, convex, becoming plane or broadly umbonate, reddish purple to violet or grayish violet; surface smooth, slimy, somewhat streaked beneath the gluten, staining purple or violet when bruised; margin inrolled when young, becoming incurved and remaining so for a long time, sometimes fringed with remnants of the partial veil.

Gills: adnexed, violet when young, becoming pale cinnamon-brown in age, staining dark violet or purple when bruised, close, broad, with 2 tiers of attenuate lamellulae.

Stalk: 2½–4 ¼″ (6.5–11 cm) long, ⅜–¾″ (1–2 cm) thick, enlarged downward forming a club-shaped bulb, solid, becoming stuffed, dry, silky, pale violet, staining violet or purple when bruised; partial veil whitish to pale violet, cobwebby.

Flesh: thick, soft, grayish violet, slowly staining dark purple when cut or bruised; odor and taste not distinctive.

Technical Features: pileotrama homogeneous beneath a thick gelatinous pellicle; **pleurocystidia** and **cheilocystidia** not differentiated; **basidia** clavate, 4-spored; **spores** 7–9 × 4.5–5 μm, ellipsoid, faintly rugose.

Spore Print: pale cinnamon-brown.

Fruiting: scattered or in groups on the ground under conifers; September-November; Pacific Northwest from British Columbia south to Oregon; especially common in the mountains.

Edibility: unknown.

Observations: The staining reaction on all parts of this mushroom is the best field character for identification. *Cortinarius euchrous* Henry, a European species, is similar but has a fruity odor and does not stain violet or purple when bruised. *Cortinarius subfoetidus* A. H. Sm. is similar but has white gills when young.

Notes: Also known as the "Purple-staining Cortinarius."

Agaricales
Cortinariaceae

Cortinarius subfoetidus A. H. Sm., *Lloydia* 7(3): 191–195. 1944.

Cap: 1½–4″ (4–10 cm) wide, broadly convex, becoming plane or broadly umbonate, bluish lavender to lilac, fading to pale pinkish tan on the disc and remaining lilac toward the margin at maturity; surface smooth, slimy when fresh, somewhat streaked beneath the gluten; margin incurved when young, often lacerated at maturity.

Gills: adnexed to slightly decurrent, whitish when young, becoming pinkish white and finally pale cinnamon-brown at maturity, not staining when bruised, close, moderately broad, with 2 tiers of attenuate lamellulae.

Stalk: 1–3½″ (2.5–9 cm) long, ⅜–1″ (1–2.5 cm) thick, nearly equal, solid, whitish above, covered below by a silky sheath that is pale lilac on the upper portion and pinkish to whitish below; partial veil pale lavender to whitish, cobwebby.

Flesh: thick, soft, bluish lavender at first, quickly becoming pinkish white, unchanging when cut or bruised; odor sickeningly sweet and unpleasant; taste not distinctive.

Technical Features: **pileotrama** homogeneous beneath a somewhat differentiated pellicle of narrow, subgelatinous hyphae; **pleurocystidia** and **cheilocystidia** not differentiated; **basidia** clavate, 4-spored; **spores** 7–10 × 5–5.5 μm, subellipsoid, faintly wrinkled.

Spore Print: pale cinnamon-brown.

Fruiting: scattered on the ground under conifers; September-November; Pacific Northwest from British Columbia south to Oregon; fairly common.

Edibility: unknown.

Observations: The distinguishing features of this species are the bright lilac to lavender colors of the cap and stalk, the sticky cap and whitish gills on young, fresh specimens and the sickeningly sweet, unpleasant odor. *Cortinarius mutabilis* is very similar but has violet gills at first. *Cortinarius balteatus* Fr. is also very similar but has larger spores (10–11 × 5.5–6 μm).

Notes: For additional information see Smith (1944).

Gymnopilus punctifolius (Pk.) Singer, *Lilloa* 22: 561. 1951.

Cap: 1–4″ (2.5–10 cm) wide, convex to nearly plane, dull purple-red with bluish green to greenish yellow, olive or brown tones; surface dry, often fibrillose over the disc when young, becoming smooth in age; margin incurved when young and often remaining so in age, entire, often wavy at maturity.

Gills: attached to sinuate when young, often deeply emarginate in age, yellowish olive when young, becoming pinkish cinnamon at maturity, close, broad, with numerous attenuate lamellulae.

Stalk: 1½–5½″ (4–14 cm) long, ¼–¾″ (7–20 mm) thick, nearly equal or sometimes enlarged near the apex or base, colored like the cap or paler, with fine streaks, becoming hollow in age.

Flesh: greenish yellow, moderately thick, firm; odor not distinctive; taste bitter.

Technical Features: pileipellis a repent zone of brown hyphae with a turf of slender, erect, filamentous to subventricose, capitate pileocystidia; **clamp connections** present; **pleurocystidia** 27–45 ×

3–5 μm, ventricose, capitate, hyaline; **cheilocystidia** similar to the pleurocystidia; **basidia** 4-spored; **spores** 4–6 × 3.5–5 μm, oval, with small warts, thin-walled, dextrinoid.

Spore Print: reddish yellow.

Fruiting: scattered, in groups or clusters on the ground or on decaying wood under conifers; August-December; reported from Washington, Oregon, California, Idaho, Wyoming, New Mexico, Michigan and Massachusetts.

Edibility: unknown.

Observations: The dull purple-red cap with green to yellow, olive, or brown tones, a similarly colored stalk which becomes hollow in age, greenish yellow flesh with a bitter taste and a reddish yellow spore print are the distinctive features. The cap and stalk colors fade quickly after picking. This mushroom closely resembles a *Cortinarius* species but lacks a partial veil.

Notes: For additional information see Hesler (1969).

Inocybe geophylla var. **lilacina** (Pk.) Gill., *Les Hyménomycètes* 520. 1876.

Cap: ⅝–1½″ (1.5–4 cm) wide, bluntly conical to bell-shaped or somewhat flattened with a broad umbo, pale to dark lilac when young, becoming pale pink to lilac-gray or tan in age, often retaining lilac tones on the disc; surface dry, silky-glossy to smooth; margin incurved and entire when young, becoming cracked and lacerated in age.

Gills: adnate to sinuate, whitish to pale lilac when young, becoming pale grayish brown at maturity, close, moderately broad, with 1–3 tiers of attenuate lamellulae.

Stalk: 1½–2½″ (4–6.5 cm) long, ¼–⅜″ (6–10 mm) thick, equal, solid, lilac when young, pale pink to pinkish brown in age, whitish and finely powdery at the apex, silky below; partial veil cobwebby, evanescent.

Flesh: whitish, thick at the disc, thinner toward the margin; odor spermatic to unpleasant; taste not distinctive or somewhat disagreeable.

Technical Features: **pileipellis** a trichodermium of radially arranged hyphae 5–7 µm in diameter with clamp connections present; **pleurocystidia** abundant, fusoid-ventricose, thick-walled, 40–58 × 10–18 µm; **cheilocystidia** abundant, similar to the pleurocystidia; **basidia** 25–34 × 6–8 µm, 4-spored; **spores** 7–9 × 4.5–5.5 µm, elliptical to oval, smooth, thin-walled.

Spore Print: brown.

Fruiting: scattered or in groups on the ground in conifer and hardwood forests; August-November; widely distributed in North America; common.

Edibility: poisonous.

Observations: The bluntly conical lilac cap, pinkish stalk and brown spore print are important features for the identification of this mushroom. Similarly colored species of *Cortinarius* are larger and have rusty brown spores.

Notes: This species is also known as the "Lilac Fiber Head." For a key to the varieties of *I. geophylla,* see Kuyper (1986).

Inocybe pudica Kühner, *Ann. Sci. Franche-Comte* 2: 26. 1947.

Cap: ¾–1¾″ (2–4.5 cm) wide, bluntly bell-shaped to convex, white; surface shiny, smooth to silky-glossy when dry, becoming lubricous when wet; margin entire when young, becoming uneven and lacerated in age.

Gills: sinuate, whitish when young, cinnamon-brown in age, often developing brownish red or dull pinkish stains, subdistant, narrow, with several tiers of attenuate lamellulae.

Stalk: 1–2¼″ (2.5–5.7 cm) long, $^{3}/_{16}$–$^{5}/_{16}$″ (4–8 mm) thick, nearly equal down to a slightly enlarged base, white, pruinose at the apex, smooth below and often sheathed at the base by a thin layer of white fibrils, solid; partial veil white, cobwebby, evanescent, evident only on very young specimens; ring absent.

Flesh: whitish, thin, fragile; odor spermatic; taste not distinctive.

Technical Features: cheilocystidia with ventricose, moderately thick-walled metuloids 45–50 × 19–22 µm, and thin-walled, clavate, hyaline cells 22–27 × 11–16 µm; **pleurocystidia** ventricose, with or without a short cylindrical neck, above a slender pedicel; **basidia** clavate, 4-spored; **spores** 7–10.5 × 5–6.5 µm, almond-shaped with a bluntly rounded apex, smooth, thin-walled.

Spore Print: brown.

Fruiting: scattered or in groups on moist soil among mosses, in conifer or mixed woods; August-December; common in the Pacific Northwest, reported from Nova Scotia and apparently rare east of the Rocky Mountains.

Edibility: unknown.

Observations: The white cap, whitish gills that become cinnamon-brown at maturity, brownish red to dull pinkish stains on the gills and the spermatic odor are distinctive features. *Inocybe geophylla* (Sowerby ex Fr.) Kumm. is a widely distributed poisonous mushroom which is nearly identical but lacks brownish red or dull pinkish stains on its gills and has a more sharply pointed cap. When dried, specimens of *I. pudica* typically show pink to reddish stains on the cap and salmon-pink stains on the stalk.

Notes: For additional information see Grund and Stuntz (1981).

Ramicola americana Horak and Miller in ed.

Cap: ⅛–½″ (4–12 mm) wide, convex, flat to broadly umbonate in age, dry, minutely granular to pruinose, pale gray, olive-brown to olive-yellow; margin minutely striate.

Gills: deeply adnexed, ventricose, distant, buff when young to olive-yellow in age, covered with small water drops when young.

Stalk: 1⁄16–3⁄16″ (1.5–4 mm) long, 1⁄16–⅛″ (1.5–3 mm) wide, curved, central or eccentric, equal, dry, white pruinose over an olivaceous buff ground color.

Flesh: soft, cream to buff; odor mild; taste not distinctive.

Technical Features: pileipellis a trichodermial palisade; **pileocystidia** 25–40 × 3–6 µm, cylindric, capitate, clamped; **cheilocystida** and **pleurocystidia** 21–60 × 3–7 µm, hypha-like with a capitate apex, clamped; **basidia** 20–25 × 7–9 µm, clavate, thin-walled, 2- to 4-spored; **spores** 7–10 × 5.5–7 µm, elliptical to ovoid, thin-walled, pale brown, with or without a weak pore, inamyloid.

Spore Print: brown often with an olivaceous tint.

Fruiting: gregarious on aspen logs in the summer, known only from Alberta, Canada, and in one location in California; rare.

Edibility: unknown.

Observations: The small, convex, pruinose, fruiting body with the curved, pruinose stalk, combined with the trichodermal pileipellis and brown, ovoid, thin-walled spores is a distinctive combination of characteristics for *Ramicola*. The long, clamped cystidia, clamped trichodermal cells, 4-spored basidia and pip-shaped to ovoid spores separate *R. americana* from the other species in the genus. It is very possible that it has been misidentified as a *Crepidotus* and may be much more common in North America than records show to date. In Europe, species of *Ramicola* have been reported on oak and birch.

Notes: *Simocybe* was described by Karsten in 1879 but the genus has not always been accepted. Species have been placed in *Naucoria* and *Crepidotus*. More recently Døssing, in *Nordic Macromycetes* (1992), recognizes the genus *Ramicola* Velen. and transfers the species of *Simocybe* to that genus. He considers *R. rubi* to be a synonym of *R. haustellaris* (Fr.: Fr.) Watling. It has been illustrated in *Flora Agaricina Danica,* Plate 125, Fig. C as *Naucoria effugiens* Quél.

Pouzarella nodospora (Atk.) Mazzer, *Bibliotheca Mycologica* 46: 122–125. 1976.

Nolanea nodospora Atk., *Jour. Mycol.* 8: 114. 1902.

Cap: ⅜–1¾″ (1–4.5 cm) wide, conical to broadly conical when young, becoming bell-shaped to nearly flat with an umbo in age, grayish brown to reddish brown; surface dry, fibrillose-scaly to scaly; margin incurved, becoming even or wavy at maturity.

Gills: deeply adnexed to nearly free, sinuate, pale gray when young, becoming grayish brown to reddish brown in age, subdistant, moderately broad, with 2–3 tiers of attenuate lamellulae.

Stalk: 1½–4″ (3.8–10 cm) long, ⅛–¼″ (3–7 mm) thick, nearly equal, covered overall with fine, short, reddish brown to grayish brown hairs, solid, typically with coarse, stiff, dark reddish brown hairs at the base.

Flesh: grayish brown, thin, soft; odor and taste not distinctive.

Technical Features: **pileipellis** hyphae 10–25 µm wide, made up of long cylindric cells with dark brown, irregularly thickened walls up to 2 µm thick; **pleurocystidia** absent; **cheilocystidia** 36–76 × 9–18 µm, clavate to clavate-mucronate, often septate at the base; **caulocystidia** highly differentiated setiform hairs with smooth, thick walls and ventricose bases, with long attenuated necks, 100–900 µm long; **basidia** mostly 4-spored; **spores** 13–16 × 7–9 µm, elliptical, angular, 6- to 8-sided, smooth, moderately thick-walled.

Spore Print: pinkish cinnamon.

Fruiting: solitary, scattered, or in groups on leaf litter or decaying wood in deciduous woods; July-October; Massachusetts to North Carolina, west to Michigan; frequent.

Edibility: unknown.

Observations: This is the largest species in the genus *Pouzarella*. The grayish brown to reddish brown scaly cap, grayish brown to reddish brown hairs on the stalk and brownish gills are the distinctive characters of this species.

Notes: It is also known as the "Hairy-stalked Entoloma."

Chroogomphus jamaicensis (Murr.) O.K. Miller, *Mycologia* 56: 539–540. 1964.

Gomphidius jamaicensis Murr., *Mycologia* 10: 69–70. 1918.

Cap: 1–3¾″ (2.5–9.5 cm) broad, convex, with or without a low, indistinct umbo, nearly plane in age, smooth, sticky to slightly viscid when moist, pale to dark vinaceous-brown, with resupinate, imbricate scales on some specimens.

Gills: subdistant, broad, arcuate, decurrent, alternate with 1 tier of lamellulae, dull flesh-yellow, pinkish yellow to gray in age.

Stalk: 1½–4″ (4–10 cm) long, ⅛–⅝″ (3–15 mm) wide, tapering toward the base, solid, fibrillose at apex from remains of the partial veil, remainder glabrous, smooth, dry, cinnamon, or pink to dark wine-red, base cinnamon to light ochraceous.

Flesh: in cap firm, pale salmon to pinkish buff; upper stalk salmon to buff, base light to dark yellow, deep purple-blue in Melzer's solution, fresh flesh red with 70% alcohol; odor not distinctive; taste mild.

Technical Features: **pileipellis** of appressed, gelatinous, inamyloid hyphae 2–4.8 µm diameter; **pileotrama** of interwoven hyphae 7–12 µm diameter, with dark amyloid walls in Melzer's solution; **cheilocystidia** and **pleurocystidia** 110–165 × 10–25 µm, fusiform to narrowly clavate, thick-walled (4–5 µm thick), walls sometimes partially amyloid; **basidia** 36–50 × 9–11 µm, clavate, 4-spored; **spores** 17–20 × 4.5–6 µm, subfusiform in profile, elliptical in face view, smooth, thin-walled; **clamp connections** absent in the fruiting body but present on the amyloid, vegetative, basal hyphae in Melzer's solution.

Spore Print: smoky gray.

Fruiting: single to several, under 2-, 3- and 5-needle pines from sea level to about 5000 feet elevation; usually found in August through October; distributed in southeastern United States from Virginia south to Florida and Jamaica, west to Illinois and Alabama; infrequent.

Edibility: unknown.

Observations: The viscid cap with a very thin gelatinous pellicle, amyloid pileotramal cells, and pigmented flesh are characters of *Chroogomphus* (Miller 1964). *Chroogomphus vinicolor* (Pk.) O. K. Miller has thicker cystidial walls (5–7.5 µm), a conspicuous umbo on the cap, and a wide distribution in North America. A drop of Melzer's solution on the fresh cap flesh will turn deep violet. Species of *Chroogomphus* are ectomycorrhizal and *C. jamaicensis* is found associated with, and appears to be an obligate mycorrhizal partner of, pines.

Chroogomphus leptocystis (Sing.) O.K. Miller, *Mycologia* 56: 533–536. 1964.

Gomphidius leptocystis Sing., *Pap. Mich. Acad. Sci.* 32: 148. 1948.

Cap: ¾–3½″ (2–9 cm) broad, convex with a low broadly conic umbo, nearly plane in age, dry, coarsely fibrillose to minutely squamulose over the disc which is dark salmon to light ochraceous-salmon, remainder a drab ground color; margin inrolled at first, with a dense fibrillose partial veil, straight in age.

Gills: decurrent, broad, subdistant, with 3 tiers of lamellulae, many forked, orange-gray to gray when young becoming darker as the spores mature.

Stalk: 2⅜–5⅛″ (6–13 cm) long, ¼–¾″ (6–20 mm) wide, narrowing downward, dry, minutely fibrillose, with a superior fibrillose annular ring just at the apex, uniformly cinnamon-buff to ochraceous-brown, fibrils reddish from handling, becoming smoky black from the spores in age.

Flesh: firm, pale orange, yellow at the base; odor not distinctive; taste slightly acidic.

Technical Features: **pileipellis** of interwoven, filamentous hyphae 4.5–10 µm diameter, hyaline in Melzer's solution, with hypha-like end cells; **pileotrama** of interwoven hyphae 10–18 µm diameter, thin-walled, with scattered amyloid hyphae in Melzer's solution; **lamellar trama** of interwoven hyphae 5.4–13.5 µm diameter, thin-walled, hyaline in 3% KOH, yellow in Melzer's solution, with scattered amyloid cells; **clamp connections** present only on the amyloid vegetative hyphae beneath the fruiting body; **cheilocystidia** and **pleurocystidia** 100–195 × 15–18 µm, cylindric to fusiform, slightly thick-walled, (1–1.5 µm thick), occasionally walls near the center of the cystidia are amyloid in Melzer's solution; **basidia** 24–52 × 6–10 µm, clavate, thin-walled, 4-spored; **spores** 12–18 × 6–7 µm, subfusiform in profile, elliptical in face view, smooth, inamyloid.

Spore Print: smoky gray.

Fruiting: single to gregarious under western white pine and western hemlock; in September and October; from about 2500–4500 feet elevation in Oregon, Washington, Idaho, and British Columbia; infrequent.

Edibility: edible.

Observations: Amyloid hyphal cells are found in the pileotrama of all species of *Chroogomphus*. However, the narrow hyphal end cells in the pileipellis, reduced amyloid tramal tissue, and only slightly thickened walls of the cystidia distinguish *C. leptocystis* from *C. tomentosus* (Murr.) O. K. Miller and *C. sibiricus* (Sing.) O. K. Miller, a European species.

Chroogomphus pseudovinicolor O.K. Miller, *Mycologia* 58: 855–861. 1966.

Cap: 2½–5″ (6.5–12 cm) broad, convex, broadly convex in age, robust; surface dry, obscurely mottled with fine patches of tomentum especially over the margin, but elsewhere as well, pale orange-ochraceous over the margin, dull orange to red overall; margin fringed at first from the remains of the fibrillose partial veil.

Gills: subdistant, decurrent, broad, forking and frequently anastomosing, orange when young, dingy ochraceous often with a tint of olive in age, becoming smoky gray from the spores.

Stalk: 2¼–3½″ (6–9 cm) long, ¾–1½″ (2–4 cm) wide, enlarged somewhat at the annular zone just below apex, gradually tapering down to a blunt base, above annular zone unpolished and orange-buff, below annular zone appressed fibrillose to squamulose; squamules gray tinted red over an orange ground color.

Flesh: in cap firm, somewhat tough in age, dull orange to orange-red; upper stalk dull orange-ochraceous, pale buff in the base; deep purple-blue with Melzer's solution, when fresh red with 70% alcohol; odor faintly of iodine; taste sweet, pleasant.

Technical Features: **pileipellis** a densely packed trichodermium of amyloid to yellow hyphae 4–13 μm diameter; **subpileipellis** with dense lipoidal material; **pileotrama** of interwoven hyphae 4–23 μm diameter, dark amyloid walls with yellowish contents in Melzer's solution; **cheilocystidia** and **pleurocystidia** numerous 88–200 × 13–20 μm, fusiform to nearly cylindric, with thick walls (5–6 μm) often amyloid at the center; **basidia** 46–67 × 11–14 μm, clavate, 4-spored, staining blue in sudan black B; **spores** 15–20 × 5–7.5 μm, subfusiform in profile, narrowly elliptical in face view; **clamp connections** absent in fruiting body, basal hyphae below the stalk amyloid with occasional clamp connections.

Spore Print: olive-green to smoky gray-brown with a greenish cast.

Fruiting: solitary to commonly caespitose in twos or threes, under or near Douglas fir and ponderosa pine; distributed in Idaho and eastern Oregon from September through November; rare.

Edibility: unknown, other species in the genus are edible.

Observations: The dry, nongelatinous cap, amyloid cells in the pileotrama, and pigmented flesh in the cap are characters of *Chroogomphus. C. pseudovinicolor* is a large, robust, often caespitose mushroom with large thick-walled (5–6 μm) cystidia, a green to brown-tinted olivaceous spore print, and a nongelatinized pileipellis with a well-developed lipoidal layer. These distinctive features are combined with its occurrence near or under Douglas fir and ponderosa pine in the central Rocky Mountains. It is most likely a mycorrhizal associate of these conifers.

Agaricales
Gomphidiaceae

Gomphidius oregonensis Pk., *Bull. Torrey Bot. Club* 25: 326. 1898.

Cap: 1–6″ (2.5–15 cm) wide, convex when young, becoming broadly convex to nearly plane in age; surface slimy, color variable, grayish cinnamon to pale salmon-buff when young, becoming dark reddish brown at the center and vinaceous-cinnamon to salmon-pink toward the margin at maturity; margin entire, incurved and often with fibrillose partial veil remnants beneath the layer of gluten, uplifted and somewhat wavy in age.

Gills: decurrent, occasionally forked, white then smoky gray-brown, subdistant to close, moderately broad, with 2–3 tiers of attenuate lamellulae.

Stalk: 2–4¾″ (5–12 cm) long, ⅜–1⅛″ (1–3 cm) thick, solid, equal, tapering downward or somewhat enlarged at the base, usually partially buried, white above, whitish below the annular zone, becoming yellow to golden yellow at the base, slimy; gluten forming pinkish cinnamon to nearly black streaks and bands;

partial veil hyaline, slimy, with a white, fibrillose veil beneath, leaving a thin, superior annular zone that becomes blackish brown from trapped spores.

Flesh: white in cap except near the pileipellis where it becomes the color of the cap, soft and spongy, thick at the disc, tapering abruptly at the margin; white in stalk, becoming yellow to golden yellow in the lower one-third to one-half; odor and taste not distinctive.

Technical Features: **pileipellis** a gelatinous pellicle of hyphae under 2 µm diameter; **pileotrama** of interwoven, thin-walled, inamyloid hyphae; **cheilocystidia** and **pleurocystidia** clavate to cylindric, thin-walled, 80–120 × 8–13 µm; **basidia** 4-spored, clavate; **spores** 10.5–13(–16) × 4–7 µm, oblong to elliptical, smooth, thin-walled.

Spore Print: blackish brown.

Fruiting: usually in clusters, occasionally solitary, on the ground under conifers; August-December; Michigan and the Pacific Northwest from British Columbia south to California, Montana and Idaho; fairly common.

Edibility: edible.

Observations: This species is separated from all other species in the genus by the presence of a veil and its tendency to grow in clusters which are often partially buried. In addition, it is distinguished by its shorter spores, a feature which separates it from other *Gomphidius* species such as *G. glutinosus* (Schaeff.: Fr.) Fr. and *G. subroseus* Kauff. which share the same host complex.

Notes: See Miller (1971, 1129–1163) for more information.

Hygrophorus borealis Pk. f. **borealis**, *N.Y. State Mus. Ann. Rept.* 26: 64. 1874.

Camarophyllus borealis (Pk.) Murr., *North American Flora* 9: 385. 1916.

Cap: ⅜–1¾″ (1–4.5 cm) wide, convex to obtusely convex, becoming nearly plane, dull white when fresh, fading to nearly chalk white when dry, typically pale yellow to tan over the disc; surface smooth, moist; margin smooth, entire, becoming wavy and translucent-striate in age.

Gills: arcuate and decurrent, white, moderately distant to distant, narrow, intervenose, with 3–4 tiers of attenuate lamellulae.

Stalk: ¾–3½″ (2–9 cm) long, ⅛–5⁄16″ (3–8 mm) thick, equal overall or tapering downward, whitish, smooth, dry, hollow or stuffed.

Flesh: whitish, moderately thick on the disc, soft; odor and taste not distinctive.

Technical Features: **pileipellis** of repent hyphae not sharply differentiated from the pileus trama; **clamp connections** present; **gill trama** interwoven; **pleurocystidia** and **cheilocystidia** absent; **basidia** 2- to 4-spored; **spores** 7–11 × 4.5–7 µm, elliptical, smooth, thin-walled, hyaline, inamyloid.

Spore Print: white.

Fruiting: scattered or in groups on soil under conifers or mixed woods; August-December; widespread in North America; common.

Edibility: edible.

Observations: *Camarophyllus niveus* (Fr.) Wünsche is nearly identical but has a more depressed cap, is sticky when moist, and has thinner flesh. Some authors believe they are the same species. Several other species are similar and are differentiated microscopically.

Notes: For additional information see Hesler and Smith (1963).

Hygrophorus caeruleus O. K. Miller, *Mycologia* 76: 816–819. 1984.

Cap: 2–3¾″ (5–9.5 cm) broad, broadly convex, felty, moist but not viscid, glabrous, rimose and areolate in age, cream color with a tint of blue especially over the margin.

Gills: close to subdistant, adnate to short decurrent, thick, broad to ventricose, lamellulae in two tiers, bluish green to blue-gray, waxy when crushed.

Stalk: 1–2″ (2.5–5 cm) long, ½–1″ (1.2–2.5 cm) wide, equal but tapering abruptly at base, dry, minutely pruinose at apex, with appressed fibrils below, bluish green to dingy light brown, cream color at apex, with fine white rhizomorphs entending from the base.

Flesh: dingy cream color tinted bluish green, bruising bluish gray to blue-green but soon fading, brownish just at the base; odor of rancid meal, strong at times; taste mild at first but soon unpleasant.

Technical Features: pileipellis a mixocutis of hyphae 2.5–5 µm diameter, hyaline in 3% KOH and Melzer's solution; **lamellar trama** of parallel, thin-walled hyphae 3.4–14.5 µm diameter;

clamp connections present throughout; **basidia** 35–45 × 7–8 µm, clavate, thin-walled, 4-spored; **spores** 6.5–9 × 4–5 µm, elliptical, thin-walled, inamyloid.

Spore Print: white

Fruiting: several together, usually partially buried in the duff, under Douglas fir, Engelmann spruce, and grand fir; in early spring, late June and July, in the mountains; known from central Idaho, eastern Oregon and Washington; rare.

Edibility: unknown.

Observations: The distinctive characters include a robust fruiting body with blue-gray to bluish green lamellae, a tinted blue cap margin and stalk, and a strong odor of rancid meal. A May 23, 1994, postcard from Buck McAdoo reported that *H. caeruleus* was found in eastern Washington in early May during the annual morel hunt. The exact location was near the Deer Creek Campground above Lake Wenatchee at 2700 feet elevation and identified by Dr. Joseph Ammirati. This is a substantial distribution west from the type locality in the Payette National Forest in central Idaho.

Hygrophorus inocybiformis A. H. Sm., *Mycologia* 36: 246. 1944.

Cap: 1⅛–2⁵⁄₁₆″ (3–6 cm) broad, conic, campanulate or convex with a low umbo, dry, innately fibrillose, with appressed squamules, dark gray to drab; margin incurved when young and fringed white at first from the remains of the partial veil.

Gills: short decurrent, subdistant, broad, thick, waxy, alternating with short, thick lamellulae, nearly white tinted gray to grayish buff.

Stalk: 1⅛–2⅜″ (3–6 cm) long, ³⁄₁₆–½″ (5–12 mm) thick, equal or tapering somewhat at the base, dry, silky white at apex above the superior fibrillose annulus, below with abundant gray-brown fibrils over the white ground color extending to the base.

Flesh: soft and fragile, tinted gray beneath the pileipellis, remainder of cap and stalk white and firm; odor none; taste mild.

Technical Features: pileipellis a trichodermium of irregular, often clustered hyphae 10–15 µm diameter which form squamules, arising from a layer of repent, nongelatinous hyphae; **lamellar trama** of divergent hyphae 6–10 µm diameter; **clamp**

connections present and common; **cystidia** absent; **basidia** 60–85 × 10–12 µm, narrowly clavate, thin-walled, 4-spored; **spores** 9–14 × 5–8 µm, elliptical, thin-walled, smooth, inamyloid.

Spore Print: white.

Fruiting: usually gregarious, on soil under spruce and fir; summer and fall; distributed in the mountains and conifer forests of the western United States including Alaska and Canada; infrequent.

Edibility: unknown

Observations: The drab squamulose dry cap, gray-brown squamules over the white ground color of the stalk, thick white decurrent gills, long narrow cystidia and the elliptical large spores are a combination of characters which distinguish this species. It looks like a species in the *Tricholoma terreum* complex but has thick, waxy, decurrent gills and much larger spores. Hesler and Smith (1963) placed it in the subsection Camarophylli, series Camarophylli. It was first described in North America and is reported from Scandinavia in *Nordic Macromycetes* (Hansen and Knudsen 1992).

Hygrophorus marzuolus (Fr.) Bres., *Atti. Acad. Agiati Rovereto* 2: 3. 1893.

Cap: 1–4¼″ (2.5–11 cm) broad, broadly convex, glabrous, viscid to waxy subviscid with a sheen, light to pale gray-brown over the center; margin darker brown to blackish gray, sulcate to sulcate-striate in older specimens.

Gills: adnate to adnexed, distant, broad, with two tiers of lamellulae, white at first, soon gray to bluish gray, unchanging or darkening slightly in age.

Stalk: 1¼–4″ (3–10 cm) long, ⅜–1″ (1–2.5 cm) wide, equal or tapering somewhat toward the base, at apex with a few tufts of fibrils, glabrous below, dry, white tinted gray over the upper half.

Flesh: conspicuously water-soaked beneath pileipellis, remainder tinted gray with a water-soaked sheen; stalk dull white with a sheen; odor none; taste mild.

Technical Features: pileipellis an ixomixocutis in a narrow band, 1.8–4.5 µm diameter, thin-walled, hyaline in 3% KOH, yellowish in Melzer's solution; **lamellar trama** of divergent hyphae 4.5–11 μm diameter, thin-walled, hyaline in 3% KOH, light yellowish in Melzer's solution; **clamp connections** present in the pileipellis; **cheilocystidia** and **pleurocystidia** absent; **basidia** 42–55 × 5.4–8.1 μm, narrowly clavate, 4-spored, occasionally 2-spored, hyaline; **spores** 6.5–8.5 × 4.5–5 μm, elliptical, thin-walled, smooth, inamyloid.

Spore Print: white.

Fruiting: single to several, sometimes caespitose, under conifers (very often subalpine fir and Engelmann spruce) in needle duff, usually in the wet zone at high elevations, often in melting snowbanks in the spring; fruiting initiated under the snowbanks in late May through July; distributed throughout the mountains in Idaho, Montana, Washington, Oregon, northern Arizona, and northern California; infrequent.

Edibility: listed as edible in Europe.

Observations: The robust fruiting body, waxy gray gills, water-soaked gray flesh and the occurrence near melting snow banks are the distinctive characters. A fall fruiting species, *H. camarophyllus* (A. and S.: Fr.) Fr., is similar but has white gills which are very short decurrent. In Europe, *H. marzuolus* is also found in similar habitats in the very early spring (Breitenbach and Kränzlin 1991).

Hygrophorus purpurascens (Fr.) Fr., *Epicr. Myc.,* 322. 1838.
Agaricus purpurascens Fr., *Sys. Myc.*, 34. 1821.

Cap: 2⅜–6″ (6–15 cm) broad, convex, plane in age, viscid, appressed fibrillose, with red to purplish red fibrils over white ground color, often appears streaked with red pigment, commonly covered with needles and debris; margin incurved until late maturity, conspicuously streaked with radially oriented, red to purplish red fibrils over a largely white ground color.

Gills: adnate at first to short decurrent in age, subdistant, narrow to medium broad in age, with an irregular tier of short lamellulae, white at first, often tinted pink or spotted red to purplish red in age.

Stalk: 1⅛–4″ (3–10 cm) long, ⅜–1″ (1–2.4 cm) wide, equal or tapered just at the base, dry, superior, partial veil remaining as a fragile fibrillose zone, surface above and below fibrillose zone streaked with red and purplish red fibrils over a white ground color, concolorous with the cap.

Flesh: firm, thick, white in cap and stalk; odor mild; taste pleasant.

Technical Features: pileipellis a very loose trichodermium which arises from a gelatinous, thick mixocutis of hyphae 2–4 μm diameter; **pileotrama** of interwoven hyphae 5–9 μm diameter, thin-walled, hyaline; **cystidia** absent; **basidia** 40–56 × 5–8 μm, narrowly clavate, thin-walled, 4-spored; **spores** 5.5–8 × 3–4.5 μm, elliptical, smooth, thin-walled, inamyloid.

Spore Print: white.

Fruiting: single to several (occasionally gregarious), in conifer duff, often partially or almost completely buried; under grand fir, Douglas fir, Engelmann spruce and other conifers in the spring and early summer in the Rocky Mountains, Cascades, and coast ranges; infrequent to rare in the late summer and fall in the northeastern and southeastern United States; distributed from coast to coast but most commonly encountered in the western United States and in adjacent western Canada; frequent.

Edibility: edible and pleasant.

Observations: The viscid cap, distinctive coloration of cap and stalk, partial veil, robust appearance, and habitat under spruce, fir, and other conifers are the distinguishing characters of *H. purpurascens*. Several species of *Hygrophorus* are similar including *H. russula* (Fr.) Quél. which has a weakly viscid to dry cap, close gills, no partial veil, and is found under hardwoods in the central and eastern United States and Canada. In addition, the smaller *H. erubescens* (Fr.) Fr. is similar but has a glutinous to viscid cap with yellowish stains especially in the flesh, subdistant gills, no partial veil, larger spores (7–11.5 × 5–6 μm), and is distributed in the eastern and western United States. These species may be found in Hesler and Smith (1963).

Agaricales
Hygrophoraceae

Hygrophorus tennesseensis A. H. Sm. and Hesler, *Lloydia* 2: 40. 1939.

Cap: 1½–4½″ (4–11.5 cm) wide, convex to nearly flat, often depressed in age, whitish overall in very young specimens, becoming dull yellowish brown over the central portion and whitish over the marginal area by maturity; surface smooth, sticky, somewhat slimy when wet; margin entire, often slightly striate, inrolled and somewhat cottony, especially when young.

Gills: attached and slightly decurrent, white, subdistant, moderately broad, with several tiers of attenuate lamellulae.

Stalk: 1½–4″ (4–10 cm) long, ⅜–¾″ (9–20 mm) thick, tapering downward or nearly equal, whitish, dry, covered with tiny cottony scales at the apex and scattered fibrils below, solid.

Flesh: white, thick, firm; odor of raw potatoes; taste bitter.

Technical Features: pileipellis of gelatinous, colorless or nearly colorless hyphae, 3–5 μm broad, repent or more or less erect and entangled, forming an ixotrichodermium; **clamp connections** present; **lamellar trama** divergent; **pleurocystidia** and **cheilocys-**

tidia absent; **basidia** mostly 4-spored; **spores** 6–9 × 4.5–6 µm, elliptical, smooth, thin-walled, hyaline, inamyloid.

Spore Print: white.

Fruiting: scattered or in groups on soil under conifers; September-February; New York to North Carolina, Tennessee, Kentucky, and California; frequent.

Edibility: unknown.

Observations: A brownish sticky cap with a whitish marginal area, white gills and stalk, an odor of raw potatoes and bitter tasting flesh are distinctive features. The sticky pileipellis peels easily to the disc. *Hygrophorus bakerensis* A. H. Sm. and Hesler is very similar but has an odor of almonds and mild tasting flesh.

Notes: For additional information see Hesler and Smith (1963).

Hygrophorus turundus var. **sphagnophilus** (Pk.) Hesler and A. H. Sm., *Sydowia* 8: 324. 1954.

Hygrophorus miniatus var. *sphagnophilus* Pk., *N. Y. State Mus. Rept.* 53 (for 1899): 856. 1901.

Cap: ⅜–1⅜″ (1–3.5 cm) broad, broadly convex, soon deeply depressed, dry, minutely floccose at least in center, usually scaly in age, tips of scales orange to orange-yellow, remainder red to scarlet at first, fading to yellow or orange-yellow in age; margin incurved at first, often scaly and wavy in age.

Gills: decurrent, distant, broad, thick and waxy, edges even, alternating with a tier of very short lamellulae, orange to yellow-orange.

Stalk: 1½–4¾″ (4–12 cm) long, 1⁄16–⅛″ (1–3 mm) wide, equal, glabrous, yellowish, orange-yellow to red, pale white at the base.

Flesh: thin, yellowish in the cap and stalk, with a hollow center in age; odor and taste mild.

Technical Features: pileipellis a loosely interwoven trichodermium of hyphae 16–21 μm diameter, with erect end cells 85–115 μm diameter; **pileotrama** of hyphae (4.5–) 9–28 μm diameter, vesiculose to filamentous, hyaline in 3% KOH, yellow-brown in Melzer's solution; **lamellar trama** nearly regular of hyphae 9–20 μm diameter; **clamp connections** present in hyphal trama; **cheilocystidia** and **pleurocystidia** 37–57 × 5–12 μm, clavate or fusiform, thin-walled, hyaline; **basidia** 40–70 × 7–12 μm, clavate, thin-walled, hyaline, 2- to 4-spored; **spores** 9–14 × 5–10 μm, elliptical, thin-walled, hyaline or pale yellow in Melzer's solution.

Spore Print: white.

Fruiting: gregarious or scattered in sphagnum bogs; in early summer through the fall; distributed throughout Canada and the northern United States; frequent to common.

Edibility: unknown.

Observations: The unusual habitat in sphagnum bogs; yellow, distant, decurrent lamellae, large spores, and orange to red cap with orange-yellow tips of the squamules are distinctive characters of *H. turundus* var. *sphagnophilus*. The closely related *H. turundus* (Fr.) Fr. var. *turundus,* widely distributed in North America and Europe, has squamules with dark brown tips over the cap center and it may be somewhat smaller but there are no other differences. Neither species has been noted as far south as California by Largent (1985).

Notes: Many authors place this species in the genus *Hygrocybe* P. Kumm. along with other small colorful species such as *Hygrocybe conica* (Fr.) P. Kumm.=*Hygrophorus conicus* (Fr.) Fr.

Hygrotrama foetens (Phill. ex Berk. and Br.) Sing., *Sydowia* 12: 221. 1958.

Camarophyllus foetens (Phill.) J. Lange, *Dansk Bot. Ark.* 4: 18. 1923.

Camarophyllopsis foetens (Phill. ex Berk. and Br.) Arnolds, *Mycotaxon* 25 (2): 639. 1986.

Cap: ⁵⁄₁₆–1⁵⁄₈″ (6–40 mm) broad, plane to convex-depressed, moist, glabrous, with a granular appearance, hygrophanous, gray-brown to cinnamon-brown on disc; margin striate with a paler coloration between the striae.

Gills: decurrent, distant, medium broad, thick, alternating with lamellulae or forked at or near the margin, waxy, brittle, light gray.

Stalk: ⁵⁄₈–1⁵⁄₈″ (1.6–4 cm) long, ¹⁄₈–³⁄₁₆″ (1.2–4 mm) wide, broad at apex tapering to a narrow base, terete or slightly flattened, obscurely but definitely banded or obscurely scabrous-dotted over the surface, brown near the apex, with a white canescence over the base.

Flesh: firm, gray-brown; odor unpleasant, of mothballs, naphthalene flakes or chloride of lime; taste mild.

Technical Features: **pileipellis** hymeniform, cells subglobose to pyriform 10–60 × 5–26 µm, thin-walled, hyaline; **pileotrama** of interwoven hyphae 3–8 µm diameter; **lamellar trama** similar to the pileotrama, interwoven; **clamp connections** absent; **cystidia** absent; **basidia** 34–42 × 5–7 µm, narrowly clavate to hyphal-like, thin-walled, 4-spored, hyaline; **spores** 4.5–6 × 4–5 µm, subglobose, thin-walled, hyaline, smooth, inamyloid.

Spore Print: white.

Fruiting: several together or caespitose, in moss or on naked soil, under hardwoods (e.g., dogwoods and tulip trees); from August through October; reported from Idaho, Michigan, and Virginia; rare.

Edibility: unknown.

Observations: The distinctive features include the small fruiting body with decurrent gills, hymeniform pileipellis, disagreeable but distinctive odor, subglobose, hyaline, inamyloid spores, and narrow basidia.

Notes: Hesler and Smith (1963) placed this taxon in the genus *Hygrophorus,* section *Hygrotrama*. Moser (1983) placed it in the genus *Hygrotrama* Sing. but more recently it has been placed in the genus *Camarophyllopsis* Herink in the *Nordic Macromycetes* (1992). The basic characteristics of each of these taxonomic placements is the same and is based on the hymeniform pileipellis, decurrent lamellae, and hyaline, globose to short elliptical spores.

Neohygrophorus angelesianus (A. H. Sm. and Hesler) Singer, *Lilloa* 22: 149. 1951.

Hygrophorus angelesianus A. H. Sm. and Hesler, *Lloydia* 5: 6. 1942.

Clitocybe mutabilis Bigelow, *Mycotaxon* 6: 181. 1977.

Cap: ¾–2″ (2–5 cm) broad, convex to nearly plane with a shallow, depressed disc in age, glabrous, moist but not viscid, brownish gray to dark gray-brown; margin incurved at first, soon straight, often striate or crenate and sometimes uplifted in age.

Gills: short to long decurrent, waxy, narrow at first to broad in age, with two tiers of lamellulae, violaceous when young to violaceous-brown or vinaceous-brown in age.

Stalk: ¾–2″ (1.8–5 cm) long, ⅛–¼″ (2–5 mm) wide, equal, glabrous when moist, when dry with cream-colored fibrils over the apex, brown with a tint of violet at the base, with a white or pale lilac-tinted mycelium; with scant white rhizomorphs at the base.

Flesh: firm, dark gray in cap and stalk; red reaction in 3% KOH on fresh flesh of gills and stalk; odor not distinctive; taste mild and bland.

Technical Features: **pileipellis** a mixocutis of hyphae 1.5–5.5 μm diameter; **pileotrama** of interwoven hyphae, 3–10 μm diameter; **lamellar trama** of subparallel hyphae 2–7 μm diameter, vinaceous-red in 3% KOH; **clamp connections** present on all tissues; **pleurocystidia** and **cheilocystidia** absent; **basidia** 33–51 × 4.5–6 μm, clavate, 4-spored, thin walled; **spores** 7–9 (–11) × 4–5.5 μm, elliptical to broadly elliptical, thin-walled, amyloid.

Spore Print: white.

Fruiting: single to several on soil and needles, in or near melting snow banks; in May, June, and July; distributed throughout Wyoming, Utah, Montana, Idaho, Washington, Oregon, and California; frequent to common at high elevations.

Edibility: unknown.

Observations: The small convex-depressed cap, decurrent, violaceous-brown, waxy gills, amyloid spores, red reaction in 3% KOH of gill and stalk tissue, and the fruiting near or in melting snow banks in the western mountains of North America are a combination of distinctive features of this unique species. It is sometimes found fruiting directly in the snow in the melt zone as illustrated by Miller (1965b).

Notes: The combination of unique features is neither typical of *Hygrophorus* nor of *Clitocybe*. The genus *Neohygrophorus* seems appropriate for this taxon.

Pluteus petasatus (Fr.) Gill., *Les Hyménomycès de France,* 395. 1874.

Cap: 3⅛–8″ (8–20 cm) broad, convex to almost plane in age, dry, glabrous with small, often appressed squamules, white with grayish tints, usually gray or grayish brown over the often areolate disc.

Gills: narrowly free, close at first, subdistant in age, broad, pure white becoming deep pink in age.

Stalk: 2⅜–3 ⅛″ (6–8 cm) long, ⁵⁄₁₆–½″ (6–12 mm) wide, equal, glabrous, dry, dull white except at the base which may have some mottled brown stains or may appear furrowed.

Flesh: firm, soft, white in cap and stalk; odor and taste not distinctive.

Technical Features: pileipellis of filamentous, hyaline, thin-walled hyphae 2.5–11 µm diameter; **pileotrama** interwoven, often inflated, thin-walled hyaline, hyphae 11–27 µm diameter; **lamellar trama** convergent, hyaline hyphae 3.4–20 µm diameter;

clamp connections absent; **pleurocystidia** and **cheilocystidia** 50–70 × 13–19 µm, frequent, protruding, thick-walled, fusiform, usually with 1–3 horns or ears; **basidia** 25–30 × 7–8 µm, clavate, thin-walled, hyaline, 4-spored; **spores** 6–8 × 4–5 µm, elliptical, hyaline, thin-walled.

Spore Print: pink.

Fruiting: on wood debris, mulch, and sawdust piles; in spring, summer and fall; distribution widespread throughout North America; infrequent.

Edibility: unknown, most species in the genus are edible.

Observations: The robust white caps with the gray-brown to brown, areolate disc, white stalk, cystidia (called metuloids) with 1–3 hooked apices, and elliptical spores are a distinctive combination of characters. *Pluteus pellitus* (Pers.: Fr.) P. Kumm. has a pure white, smooth cap and pleurocystidia with horns. *Pluteus tomentosulus* Pk. has a pure white, velvety cap and pleurocystidia that lack horns.

Lactarius alachuanus var. **alachuanus** Murr., *Mycologia* 30: 360. 1938.

Cap: 2–3″ (5–7.5 cm) wide, convex to nearly plane, often slightly depressed and sometimes with a slight umbo, pinkish cinnamon when young, pinkish buff in age; surface azonate, smooth, viscid when wet; margin entire.

Gills: adnexed to adnate or slightly decurrent, pinkish buff, slowly darkening when bruised, close, moderately broad, sometimes forking, with several tiers of attenuate lamellulae.

Stalk: 1–2¾″ (2.5–7 cm) long, ⅜–⅝″ (1–1.6 cm) thick, nearly equal or tapering toward the base, pinkish buff, with a white canescence, pinkish cinnamon when rubbed, often curved at the base, solid.

Flesh: whitish, moderately thick, firm; odor not distinctive or sometimes aromatic; taste slightly bitter then moderately acrid.

Latex: white, unchanging, not staining the gills, slowly and moderately acrid.

Technical Features: **pileipellis** an ixolattice; **pileotrama** with clusters of spherical cells and lactifers with pale yellowish contents; **pleurocystidia** 60–72 × 6–7.5 µm, acuminate, with a subapical constriction; **cheilocystidia** 25–30 × 4–6 µm, ventricose with an obtuse apex, rare; **basidia** 4-spored; **spores** 7.5–9 × 6–7.5 µm, broadly ellipsoid; plage distinct, amyloid; ornamentation a partial to broken reticulum of nodulose ridges and isolated warts; prominences 0.8–1.5 µm high, amyloid.

Spore Print: white to creamy white.

Fruiting: scattered or in groups on sandy soil and rotting wood in mixed oak and pine woods; October-February; Tennessee and North Carolina south to Alabama and Florida; infrequent to common.

Edibility: unknown.

Observations: The distinctive features include a pinkish cinnamon to pinkish buff smooth cap, pinkish buff gills, a pinkish buff stalk with a white canescence which stains pinkish cinnamon when rubbed, and white, unchanging latex which tastes slowly and moderately acrid. Large fruitings are very common in central Florida following rainy periods. Considerable variation in cap color may be observed depending upon moisture content. *Lactarius alachuanus* var. *amarissimus* (Murr.) Hesler and A. H. Sm., a questionable variety, has very bitter flesh that is not acrid and sparse watery latex that is assumed not to taste acrid.

Notes: For additional information see Hesler and Smith (1979, 463–465).

Agaricales
Strophariaceae

Pholiota astragalina (Fr.) Singer, *Agaricales in Modern Taxonomy* 516. 1951.

Cap: ¾–2⅛″ (2–5.5 cm) wide, broadly cone-shaped to broadly bell-shaped with an obtuse umbo, bright pinkish orange, fading to yellowish orange especially at the center, sometimes developing dark brown areas of discoloration in age; surface smooth, sticky to slimy when wet; margin entire, often wavy at maturity, coated with a layer of pale yellow fibrils.

Gills: notched to nearly free, orange-yellow, slowly staining yellow-brown where bruised, close, moderately broad, with 2–3 tiers of attenuate lamellulae.

Stalk: 2–4¾″ (5–12 cm) long, 3/16–⅜″ (5–10 mm) thick, nearly equal overall, pale yellow, typically slightly fibrillose but some portions appear smooth, hollow, developing brown stains where handled; partial veil pale yellow, fibrillose; ring typically absent.

Flesh: orange to yellow-orange in cap, watery, moderately thick, firm; in stalk whitish at the apex and dull orange-tan in the base; odor not distinctive; taste bitter.

Technical Features: **pileipellis** a gelatinous pellicle of hyaline to yellow interwoven hyphae, 2–5 µm in diameter; **clamp connections** present; **pleurocystidia** 36–60 × 8–15 µm, fusoid-ventricose to clavate-mucronate, thin-walled; **cheilocystidia** 35–73 × 3–8 µm, ventricose at base, neck long, cylindric, with an obtuse apex; **basidia** 4-spored; **spores** 5–7 × 4–4.5 µm, oval to elliptical, smooth, thin-walled, with a minute germ pore.

Spore Print: reddish brown.

Fruiting: scattered or in groups on decaying, often moss-covered, conifer wood; August-October; widespread in North America but most common in the Pacific Northwest; occasional.

Edibility: unknown.

Observations: The bright pinkish orange cap and bitter tasting flesh make this mushroom easy to identify.

Notes: This species has been reported from several other countries including Canada, France, Sweden, and Switzerland. For additional information see Smith and Hesler (1968, 326–327).

Pholiota flavida var. **flavida** (Fr.) Singer, *Lilloa* 22: 516. 1949.

Cap: 1–2¾″ (2.5–7 cm) wide, convex to nearly plane, brownish orange when young, fading to brownish yellow or orange-yellow, darkest over the disc; surface smooth and slightly sticky; margin entire, sometimes wavy at maturity, typically incurved at first and often remaining so at maturity, decorated with delicate, pale yellow veil fibrils.

Gills: attached or notched, whitish to pale yellow when young, pale rusty brown when mature, close, moderately broad, with several tiers of attenuate lamellulae.

Stalk: 2–4¾″ (5–12 cm) long, ¼–⅝″ (5–16 mm) thick, nearly equal overall, solid, whitish to pale yellow above the ring, rusty brown below to the base, typically fibrillose overall except somewhat silky near the apex; partial veil pale yellow, fibrillose; ring consisting of a yellowish, fibrillose, inconspicuous annular zone.

Flesh: pale yellow, thick, firm; odor of citrus, especially grapefruit, sometimes only faintly fragrant; taste not distinctive.

Technical Features: pileipellis a thick gelatinous layer of hyphae 2.5–6 µm in diameter; **clamp connections** present; **pleurocystidia** absent; **cheilocystidia** 26–42 × 3–10 µm, variable, subclavate, subfusoid to near cylindric but mostly irregular, thin-walled, smooth; **basidia** mostly 4-spored; **spores** 7–9 × 3.5–5 µm, oval to elliptical, smooth, with a distinct apical pore, often somewhat truncate, slightly dextrinoid.

Spore Print: rusty brown.

Fruiting: in clusters or groups on the ground, often among mosses, under conifers; August-October; Pacific Northwest; occasional to locally common, especially in the Rocky Mountain area.

Edibility: unknown.

Observations: The distinctive features include a sticky, brownish orange to orange-yellow cap, a yellowish partial veil on young specimens, a stalk that is rusty brown on the lower portion, whitish above an annular zone and pale yellow flesh with a citrus-like odor resembling grapefruit. The ring is often absent or consists of a few inconspicuous fibrils. *Pholiota malicola* var. *macropoda* A. H. Sm. and Hesler is similar but has larger spores and flesh with an odor suggestive of green corn. *Pholiota malicola* var. *malicola* (Kauffm.) A. H. Sm. lacks a distinctive odor and has larger spores (8.5–12 × 4.5–6 µm).

Notes: For additional information see Smith and Hesler (1968, 177–179).

Calocybe onychina (Fr.) Donk, *Nova Hedwegia* 5: 43. 1962.
Tricholoma onychina Fr., *Epicr.* No. 146, 41. 1836.

Cap: ¾–3½″ (2–9 cm) broad, broadly convex to obscurely umbonate, nearly plane in age, glabrous, purple-red to wine-red when young, with darker shades of purple-red in age; margin inrolled at first, often crenulate in age.

Gills: adnate, close, medium broad, with one tier of lamellulae, dull deep yellow.

Stalk: ¾–1¾″ (2–4.5 cm) long, 3/16–⅝″ (4–15 mm) wide, enlarging somewhat toward the base, dry, canescent, dull white with a pinkish to purplish hue especially over the upper half.

Flesh: firm, white to pale yellow and solid throughout; odor farinaceous, even somewhat cucumber-like; taste mild.

Technical Features: pileipellis a loosely arranged hymenoderm of erect pileocystidia 20–34 × 2.7–7.8 µm, hypha-like to clavate or sometimes ten-pin shaped, thin-walled, hyaline; **pileotrama** of interwoven hyphae 4.2–6.7 µm diameter, filamentous, hyaline, thin-walled; **lamellar trama** of parallel hyphae 2.5–5.9 µm diameter, hyaline; **cystidia** absent; **clamp connections** absent; **basidia** 24–30 × 5–8 µm, clavate, thin-walled, 4-spored, with siderophilous granulations in carmine-acetic acid; **spores** 3–4 × 2–3.5 µm, subglobose to short-elliptical, thin-walled, inamyloid.

Spore Print: white.

Fruiting: several to gregarious in conifer duff, under mixed conifers including Engelmann spruce, subalpine fir, and Douglas fir; in June and early July; distributed in the central Rocky Mountains of Idaho; rare.

Edibility: edible [according to Cetto, No. 1025 (1979)].

Observations: *Calocybe onychina* is a species rarely encountered in boreal coniferous forests in Europe and North America. In western North America it is typically found associated with spruce. The purple-red to wine-red cap, bright yellow gills and the purplish to pinkish stalk are distinctive characteristics in the genus *Calocybe.* The distribution is not known but is most likely confined to the subalpine conifer forests of the western United States and Europe.

Notes: The species was first described by Fries and placed in the genus *Tricholoma*. Donk (1962) transferred it to *Calocybe* following a positive test for siderophilous granulations in the basidia.

Agaricales
Tricholomataceae

Chrysomphalina aurantiaca (Pk.) Redhead, *Acta Mycologica Sinica Suppl.* 1: 297–304. 1986.

Omphalia aurantiaca Pk., *Bull. Torrey Bot. Club.* 25: 323. 1898.

Omphalina luteicolor Murr., *North Am. Flora* 9: 348. 1916.

Cap: ⅜–1½″ (1–4 cm) wide, broadly convex to nearly plane, often slightly depressed, orange when young, fading to orange-yellow then yellow; surface silky-fibrillose at first, becoming nearly smooth at maturity, moist and hygrophanous when fresh; margin incurved when young, becoming wavy and often uplifted in age, coated with tiny whitish hairs which disappear as specimens mature.

Gills: adnate to slightly decurrent when young, becoming moderately to strongly decurrent in age, orange to orange-yellow, fading to pale yellow, distant, narrow, arched, often forked, typically intervenose, waxy-appearing, with several tiers of attenuate lamellulae.

Stalk: ⅜–1⅛″ (1–3 cm) long, 1/16–3/16 (1.5–5 mm) thick, nearly equal or tapering at either end, orange to orange-yellow when young, fading to yellow at maturity, smooth, solid then stuffed and often hollow in age, sometimes coated with a whitish basal mycelium.

Flesh: orange to orange-yellow, fading to pale yellow; odor and taste not distinctive.

Technical Features: pileipellis of interwoven cylindrical hyphae 4.5–8 µm in diameter; **clamp connections** absent; **cystidia** not differentiated; **basidia** mostly 4-spored; **spores** 7–10 × 4–5.5 µm, elliptical to obovate, smooth, hyaline, inamyloid.

Spore Print: white to pale yellow.

Fruiting: scattered, in groups or small clusters on decaying conifer wood and adjacent duff; September-December; British Columbia to California, Idaho and Michigan; fairly common.

Edibility: unknown.

Chrysomphalina aurantiaca (immature specimens)

Chrysomphalina aurantiaca (mature specimens)

Observations: The combination of orange cap and stalk colors, and tiny whitish hairs on the cap margin of young specimens, is diagnostic for this mushroom.

Notes: For additional information see Murrill (1916), and Smith (1949, 317–318).

Clitocybe inornata ssp. **occidentalis** Bigelow, *North American Species of Clitocybe, Part 1,* 188–189. 1982.

Cap: 1–3⅛″ (2.5–8 cm) wide, convex to nearly flat, pale creamy gray to pale pinkish gray when young, becoming pale pinkish brown to pale grayish brown or darker yellow-brown over the disc in age; surface dry, covered by a layer of tiny fibrils and often areolate at maturity; margin inrolled at first then incurved and often remaining so to maturity, entire, sometimes wavy in older specimens.

Gills: adnate or slightly decurrent, pale grayish brown, darkening and sometimes developing dull yellowish spots in age, close, moderately broad, arched, sometimes forked, typically intervenose, with several tiers of attenuate lamellulae.

Stalk: 1⅛–3⅜″ (2.8–8.5 cm) long, ¼–⅝″ (6–15 mm) thick, nearly equal overall, solid, whitish to pale creamy gray, developing yellow-brown areas when handled or in age, covered with tiny fibers; base often embedded in the substrate with a dense whitish to pale creamy gray tomentum.

Flesh: whitish, thick, soft; odor not distinctive; taste mild or slightly bitter.

Technical Features: pileipellis rather thick, of cylindric hyphae 2.5–4.5 μm in diameter; **clamp connections** present; **basidia** 21–40 × 5–7 (–10) μm, mostly 4-spored; **spores** 6.5–10.5 × 3–4 μm, elliptical, smooth, thin-walled, inamyloid.

Spore Print: white.

Fruiting: solitary, scattered, in groups or clusters on conifer needles or hardwood leaves; July-November; widespread in North America including Alaska, Washington, Idaho, Colorado, New Mexico, Michigan, North Carolina, New York, and Ontario; frequent.

Edibility: unknown.

Observations: The combination of a grayish cap which darkens to brown over the disc in age, a whitish stalk which stains yellow-brown when handled, the absence of a distinctive odor, a white spore print and growth on conifer needles or hardwood leaves is distinctive. *Clitocybe inornata* ssp. *inornata* (Sowerby: Fr.) Gill. is found in Europe and is nearly identical but has an odor which has been variously described as: somewhat disagreeable, rancid, like caraway seeds, or radish-like. *Clitocybe alexandri* (Gill.) Konrad, found in Europe, is also similar but has a sooty gray, larger cap (4–6″ [10–15 cm]), commonly forked gills, and smaller spores (5.5–6.5 × 3.5–4.5 μm).

Notes: *Clitocybe inornata* ssp. *occidentalis* has also been reported from Denmark, England, and France. For additional information see Bigelow (1982).

Cystoderma arcticum Harmaja, *Karstenia* 24: 31. 1984.

Cap: ⅜–2″ (1–5 cm) broad, subconic to obtuse at first, convex to broadly convex in age, often with a broad umbo, surface radially wrinkled, granulose especially over the disc, orange-brown to yellow-brown, umbo sometimes darker; margin yellow-brown with white to buff appendiculate remains of the universal veil.

Gills: close to crowded, sinuate to adnexed, broad, uneven, thick, alternating with one tier of lamellulae, pale yellow to light yellow.

Stalk: ¾–2¾″ (1.8–7 cm) long, ⅛–7⁄16″ (3–11 mm) wide, equal, tapering or enlarging toward the base, dry, with a persistent, usually flaring annulus, at the apex smooth or with downy fibrils, below the annulus granulose or with scattered fibrils, concolorous with the cap.

Flesh: firm, white in both cap and stalk, becoming brownish yellow in age; odor farinaceous; taste unknown.

Technical Features: pileipellis a polycystoderm of globose, pyriform to broadly clavate cells, 11–65 × 11–34 µm, with rusty brown walls in 3% KOH; **lamellar trama** of parallel equal to swollen hyphae 3.4–14 µm diameter; **cystidia** none; **basidia** 27–38 × 3.7–5.1 µm, clavate, thin-walled, 4-spored with a basal clamp; **spores** 5.5–8.7 × 3.7–5.1 µm, elliptic, smooth, thin-walled, amyloid in Melzer's solution.

Spore Print: pale yellow.

Fruiting: in large or small caespitose clusters in exposed organic peat, often on exposed high center polygons or peat mounds along streams in the Arctic coastal plain, occasionally among mosses; during August and early September; distributed in the North American and European Arctic but most likely worldwide in the Arctic; frequent.

Edibility: unknown.

Observations: The caespitose habit, squatty robust appearance, orange to yellow-brown coloration, small amyloid spores, flaring annulus, and Arctic distribution are a combination of characters which distinguish this taxon. It is quite common near Barrow, Alaska in tundra habitats during peak fruiting years.

Notes: This species was placed under several names until Harmaja (1979) recognized it as a separate species. It was described from North America by Miller (1993).

Laccaria amethystina Cooke, *Grevillea* 12: 70. 1884.

Cap: 3/8–2″ (1–5 cm) wide, convex to nearly plane, sometimes depressed, bright grayish purple when young and fresh, soon fading to pinkish purple then grayish to buff; surface often translucent-striate to striate, finely pruinose, strongly hygrophanous; margin incurved when young, becoming plane, wavy and often eroded in age.

Gills: arcuate to sinuate, thick, somewhat distant, dark grayish purple fading to pinkish purple, retaining the purple color longer than the cap, with 2–3 tiers of attenuate lamellulae.

Stalk: 3/4–3″ (2–7.5 cm) long, 1/16–5/16″ (1.5–8mm) thick, nearly equal or slightly bulbous at the base, longitudinally striate, fibrillose, dry, grayish purple when young, soon fading to pinkish purple to pale pinkish brown or paler; basal mycelium violet becoming white.

Flesh: pale pinkish purple, thin; odor and taste not distinctive.

Technical Features: pileipellis of interwoven hyphae with scattered fascicles of more or less perpendicular hyphae; **pleurocystidia** absent; **cheilocystidia** 26–72 × 4–12 µm, variable, from filamentous to clavate or ventricose-rostrate, thin-walled, hyaline; **basidia** mostly 4-spored; **spores** 7–10 × 6.5–10 µm, globose, hyaline, thin-walled, spiny; spines 1.5–3 µm long.

Spore Print: white to very pale violet.

Fruiting:; solitary, scattered or in groups on the ground in hardwoods or mixed woods, usually under oak or beech; August-November; widely distributed in eastern North America; frequent.

Edibility: unknown.

Observations: The combination of small size, purple colors that fade to buff, and globose spiny spores is distinctive.

Notes: For comparisons of closely related species, see Mueller (1992).

Agaricales
Tricholomataceae

Laccaria bicolor (Maire) Orton, *Trans. Brit. Mycol. Soc.* 43: 177. 1960.

Cap: ⅜–2¾″ (1–7 cm) wide, convex to nearly plane, pinkish brown to pale yellow-brown; surface dry, fibrillose to somewhat scaly, hygrophanous; margin decurved, entire when young, becoming wavy and somewhat eroded in age.

Gills: attached and slightly short-decurrent, close to somewhat distant, moderately broad, purplish to vinaceous when young, fading to pale pinkish flesh color in age, with several tiers of attenuate lamellulae.

Stalk: 1–4¾″ (2.5–12 cm) long, ⅛–5⁄16″ (3–12 mm) thick, nearly equal or slightly bulbous at the base, longitudinally striate, fibrillose, dry, pinkish brown to dark purple-brown, basal mycelium copious, lilac fading to whitish in age.

Flesh: pale pink to dull pink, thin, soft; odor and taste not distinctive.

Technical Features: pileipellis of interwoven hyphae with many large fascicles of more or less perpendicular hyphae; **pleurocystidia** absent; **cheilocystidia** 25–56 × 7–12 µm, variable from irregular to subclavate, thin-walled, hyaline; **basidia** mostly 4-spored; **spores** 6–10 × 6–9 µm, nearly globose or broadly elliptical, hyaline, thin-walled, spiny; spines 1–2 µm long.

Spore Print: white.

Fruiting: scattered or in groups, sometimes clustered on the ground under conifers and mixed woods; August-November; widely distributed in North America; very common in the West but less so in the East.

Edibility: edible, but of poor quality.

Observations: The pinkish brown to pale yellowish brown dry, fibrillose cap, the purplish to vinaceous color of young gills, and a longitudinally striate stalk with a copious lilac basal mycelium are the distinctive features. Following rainy periods, large fruitings were seen in oak-pine woods in central Florida and represent a range extension for this species.

Notes: For comparisons of closely related species, see Mueller (1992).

Leucopholiota decorosa (Pk.) O. K. Miller, Volk and Bessette stat. nov., in ed.

Armillaria decorosa (Pk.) A. H. Sm. and Walters, *Mycologia* 39: 622. 1947.

Cap: 1–2⅜″ (2.5–6 cm) wide, hemispherical when very young, becoming broadly convex and nearly plane at maturity; surface dry, covered with numerous rusty brown, pointed, recurved scales; margin incurved and often remaining so at maturity, typically uneven and coated with coarse rusty brown fibers.

Gills: adnexed, white, edges finely scalloped, close, moderately broad, with several tiers of attenuate lamellulae.

Stalk: 1–2¾″ (2.5–7 cm) long, ¼–½″ (6–12 mm) thick, solid, equal or tapering upward, white at the apex, sheathed up to an annular zone with rusty brown, pointed, recurved scales and coarse fibers; partial veil flaring upward at first, consisting of coarse rusty brown fibers.

Flesh: white, firm, moderately thick; odor not distinctive; taste mild or somewhat bitter.

Technical Features: **pileipellis** consisting of very broad filamentous hyphae up to 24 µm in diameter; **cheilocystidia** 19–24 × 3–5 µm, clavate with a blunt apex, to fusiform, thin-walled, hyaline; **clamp connections** present; **basidia** 21–24 × 5.3–5.8 µm, 4-spored; **spores** 5–6 × 3.5–4 µm, ellipsoid, smooth, thin-walled, amyloid.

Spore Print: white.

Fruiting: scattered, in groups, or clustered on decaying wood, typically hardwood; September-November; Maine to North Carolina, west to Michigan and Tennessee; uncommon.

Edibility: unknown.

Observations: A very handsome mushroom. The combination of pointed, rusty brown recurved scales on the cap and stalk, white gills, a white spore print, clavate cheilocystidia, and growth on decaying wood makes this species easy to identify. *Cystoderma* species are very similar but have spherical cells in the pileipellis and partial veil, and typically grow on soil. *Armillaria* species have inamyloid spores and rhizomorphs. *Floccularia* species lack the unique pileipellis and are terrestrial, mycorrhizal fungi. Some *Pholiota* species often appear similar but have a brown spore print.

Notes: For additional information see Peck (1873) and Singer (1943).

Melanoleuca angelesiana A. H. Sm., *Mycologia* 36: 252. 1944.

Melanoleuca graminicola (Velen.) Kühner and Maire sensu O.K. Miller, *Mycologia* 69: 928–931. 1977.

Cap: ¾–4¼″ (2–11 cm) broad, broadly convex, sometimes with a small to large umbo, occasionally depressed around the umbo, moist to subviscid, glabrous or occasionally with a few appressed squamules, dark umber-brown to dark gray-brown or dark copper-brown; margin often somewhat upturned in age.

Gills: adnate to emarginate, close to subdistant, narrow, white to cream-buff or tinted gray, pinkish in age.

Stalk: ⅝–3″ (1.5–7.5 cm) long, ⅛–⅝″ (3–15 mm) wide, terete or slightly flattened, sometimes flaring just at the apex, glabrous to finely pubescent, longitudinally striate, white just at apex, darkening to brown below, usually concolorous with the cap, base often covered with a fine white mycelium.

Flesh: in cap firm, white to slightly grayish; in stalk firm, fibrous, white to dull white; odor not distinctive; taste mild to somewhat disagreeable.

Technical Features: **pileipellis** of interwoven, thin-walled hyphae, 3.4-16 µm diameter, smooth to roughened walls with small rounded projections and erect end cells, hyaline to dark brown in 3% KOH, yellow to red-brown often with granular contents in Melzer's solution; **pileotrama** of loosely interwoven hyphae 3-15 (-35) µm diameter, thin-walled, hyaline; **lamellar trama** of parallel hyphae 3-20 µm diameter, thin-walled, hyaline; **clamp connections** absent; **cystidia** absent; **basidia** 20–50 × 6–13 µm, clavate, hyaline, 4-spored; **spores** 7.5–10 × 5–6 µm, elliptical, verrucose with a suprahilar plage, with prominent amyloid warts in Melzer's solution.

Spore Print: white to buff or cream.

Fruiting: single to gregarious, occasionally caespitose on ground, near melting snow banks under conifers; in late May to mid-July, or in early August in a heavy snow pack year, at high elevations; distributed from Montana, Idaho, Wyoming to Utah, west to California, Oregon, Washington, north to Alberta, and most likely British Columbia; infrequent.

Edibility: unknown.

Observations: Further study of *M. graminicola* described by Velenovský in 1920 has led to the conclusion that it is not the high elevation species as indicated by Gillman and Miller (1977) but a small European species in grassy habitats. The description of *M. angelesiana* A. H. Smith fits our material in every way.

Mycena pelianthina (Fr.) Quél., *Champ. Jura et Vosges*, 102. 1872.
Agaricus pelianthinus Fr., *Syst. Myc.* 1: 112. 1821.

Cap: ⅝–2″ (1.5–5 cm) broad, convex to broadly convex and plane in age, glabrous, smooth, moist, hygrophanous, white tinted lilac to sordid purplish, becoming livid, gray in age; margin striate and somewhat ragged in age.

Gills: adnate, close, broad, with a tier of lamellulae, vinaceous-gray to purplish with dark purple edges.

Stalk: 1⅜–3⅛″ (3–8 cm) long, ⅛–5⁄16″ (2–6 mm) wide, equal or enlarging somewhat toward base or apex, nearly glabrous just at apex, remainder finely fibrillose, tinted purplish to violet, with scurfy white mycelium over the lower one-third and white-strigose at the base.

Flesh: thin, white, watery, tinted brown when moist; odor radish-like; taste mild with a slight radish-like flavor.

Technical Features: pileipellis a narrow layer of parallel hyphae; **pileotrama** and **lamellar trama** of hyphae 6–8 µm diameter, bright vinaceous-red in Melzer's solution; **clamp connections**

present; **pleurocystidia** and **cheilocystidia** 45–64 × 9–15 µm, fusoid-ventricose, thin-walled, smooth, apex acute; **basidia** clavate, 4-spored, thin-walled; **spores** 5.5–9 × 3–3.5 (–5) µm, elliptical, thin-walled, amyloid.

Spore Print: white.

Fruiting: single to gregarious on hardwood litter in hardwood or mixed hardwood-conifer forests; widely distributed in North America; fruiting in the summer and fall; frequent.

Edibility: nonpoisonous.

Observations: The sordid purplish to lilac-tinted, hygrophanous cap, vinaceous-gray gills with dark purple edges, white flesh without any yellow coloration, and small spores are a distinctive combination of features of *M. pelianthina*. *Mycena rutilantiformis* Murr. (Smith et al. 1979) is similar but has yellow flesh in the apex of the stalk and larger spores (8–10 × 3.5–5 µm).

Mycena semivestipes (Pk.) A. H. Sm., *North American Species of Mycena*, Univ. of Michigan Press, Ann Arbor, 521 pp, 1947.

Omphalina semivestipes Pk., *Bull. Torr. Bot. Club* 22: 200. 1895.

Cap: ⅜–1⅜″ (8–35 mm) broad, convex, convex-umbonate to plane at maturity, smooth, glabrous, slightly lubricous, deep fuscous to dark brown over the disc; margin translucent-striate and light brown.

Gills: adnate or with a decurrent tooth, subdistant, moderately broad, 2–3 tiers of lamellulae, white to sordid pink in age.

Stalk: ⅝–2⅜″ (1.6–6 cm) long, 1⁄16–3⁄16″ (1–3 mm) wide, terete, equal, glabrous or lightly pruinose, with white, strigose hairs over the base, white at the apex to light brown and darker brown over the base.

Flesh: firm, tough, cartilaginous, concolorous with the cap; odor faintly to strongly nitrous but soon disappearing or absent in buttons; taste mild to somewhat bitter.

Technical Features: pileipellis an ixomixocutis with a thin gelatinized layer of hyaline hyphae, 2.7–5.4 µm diameter; **pileotrama** of interwoven hyphae 3.6–6.3 µm diameter, light vinaceous-red in Melzer's solution; **lamellar trama** of irregular hyphae 3.6–6.3 µm diameter, vinaceous-red in Melzer's solution; **pleurocystidia** and **cheilocystidia** 24–34 × 5–11 µm, clavate, embedded among basidia, fusoid-ventricose, subcapitate or with knob-like processes, thin-walled, hyaline; **basidia** 21–30 × 4.5–5.4 µm, narrowly clavate, hyaline, 4-spored; **spores** (4.5–) 5–7 × 3–3.4 µm, short ellipsoid, thin-walled, amyloid.

Spore Print: white.

Fruiting: in dense, caespitose clusters, often several clusters together on hardwood logs and stumps; in the late fall and winter; distributed from Newfoundland and Ontario south to Tennessee and west to Missouri; frequent.

Edibility: unknown.

Observations: The dense, caespitose clusters, small spores, distinctive cystidia, vinaceous-red trama, and fall-winter fruiting pattern are a distinctive combination of characteristics which distinguish this taxon. The cheilocystidia are infrequent but more numerous than the pleurocystidia and hard to observe. This fungus has been found fruiting in November in Maryland and in mid-February in Virginia during a warm period.

Notes: Smith (1937) previously identified the mushroom as *Mycena tintinnabulum* (Fr.) Quél. but gives a thorough account of it in his monograph (Smith 1947).

Tricholoma pullum Ovrebo, *Can. Jour. Bot.* 67: 3139–3140. 1989.

Cap: 1¾–6″ (4.5–15 cm) wide, obtusely conic or convex with an incurved margin, becoming broadly convex to plane in age; surface dry, densely fibrillose over the disc, with radially arranged fibrils extending toward a smooth to sometimes squamulose margin, virgate, dark gray, typically paler toward the edge and between the fibrils.

Gills: sinuate, narrowly attached, moderately broad, often seceding in age, pale pinkish gray to grayish buff, becoming grayish buff with dull yellow tints in age, with dark gray coloration on portions of the edges and faces, entire, close, with numerous attenuate lamellulae which do not occur in distinct tiers.

Stalk: 1¼–3⅛″ (3–8 cm) long, ⅜–1″ (1–2.5 cm) thick, nearly equal overall, smooth to somewhat silky, whitish, sometimes with yellowish tints in age.

Flesh: grayish white in cap, thick, firm, whitish in stalk; odor not distinctive; taste bitter or acrid.

Technical Features: pileipellis of interwoven hyphae; **cheilocystidia** 25–35 × 8–15 µm, cylindric to broadly clavate, smooth, hyaline to dark gray; **caulocystidia** 25–35 × 6.5–11 µm, cylindric to clavate, smooth, hyaline, singly or in clusters; **basidia** 4-spored; **spores** 6.5–8 × 5–6 µm, broadly elliptic, smooth, hyaline, inamyloid; addition of PDAB produces a bright pink response on all parts.

Spore Print: white.

Fruiting: scattered or in groups on the ground under hardwoods, especially beech and oak; northeastern North America; September-November; uncommon.

Edibility: unknown.

Observations: The dark gray, virgate cap, along with the bitter or acrid taste and dark gray tints on the edges of the gills are distinctive. *Tricholoma sciodes* (Secr.) Martin described from Europe is nearly identical, if not the same species. *Tricholoma virgatum* (Fr.: Fr.) P. Kumm. is a common and widely distributed species in North America which is also similar, but has a sharply pointed umbo, grows in conifer and mixed woods and lacks the dark gray tints on the gill edges.

Notes: For additional information see Ovrebo (1989).

Xeromphalina brunneola O. K. Miller, *Mycologia* 60: 156–188. 1968.

Cap: ¼–⅝″ (6–15 mm) broad, convex-depressed to nearly plano-depressed, glabrous, moist, dull orange to reddish brown overall; margin opaque when moist, striate when drying.

Gills: decurrent, close, narrow, sparsely intervenose, dull orange-buff.

Stalk: 1⅛–2⅜″ (3–6 cm) long, ¹⁄₁₆–⅛″ (1.5–3 mm) wide, terete, tapering somewhat toward the base, dry, glabrous and orange-buff at apex, darker rusty brown downward, base nearly bulbous and covered with ochraceous pubescence, with dull white rhizomorphs at the base.

Flesh: thin, firm, brownish, cartilaginous, reddish brown in 3% KOH; odor and taste disagreeable.

Technical Features: **pileipellis** of radially arranged, rectangular or irregularly shaped cells, 5–12 μm diameter, dark red-brown in Melzer's solution and 3% KOH; **pileotrama** of interwoven hyphae, 5.5–11 μm diameter, yellow-brown in 3% KOH; **pileocystidia** 50–60 × 16–20 μm, clavate to broadly fusiform, yellow-brown in 3% KOH; **lamellar trama** of irregular to vesiculose cells 3.5–11 μm diameter, yellow-brown in 3% KOH; **clamp connections** common throughout; **pleurocystidia** and **cheilocystidia** 44–82 × 5.5–8.5 μm, subcylindric to fusiform, thin-walled, hyaline; **basidia** 17–25 × 4.2–5.5 μm, clavate, thin-walled, 4-spored; **spores** 5.5–6.6 × 2.5–3 μm, narrowly elliptical to cylindric, thin-walled, amyloid.

Spore Print: white.

Fruiting: in dense caespitose clusters on decorticated conifer logs; in late summer and fall; widely distributed across Canada and the northern United States; frequent.

Edibility: inedible.

Observations: *Xeromphalina brunneola* differs from *X. campanella* (Batsch: Fr.) Kühner and Maire by having a darker reddish brown cap and a disagreeable odor and taste, and also differs, according to Redhead (1987), by the smaller, narrowly elliptical to cylindric spores. Redhead also indicates that the spores may be allantoid but we have not observed this and he does not illustrate typical allantoid spores. The distribution of the species was greatly increased by Redhead, especially in Canada, by a close comparison of the spore morphology of the two species.

Xeromphalina kauffmanii A. H. Sm., *Pap. Mich. Acad. Sci. I.* 38: 81–82. 1953.

Cap: ¼–⅝″ (4–15 mm) broad, broadly convex to convex-depressed at maturity, glabrous, moist but not viscid, bright rusty orange at first, becoming bright orange or cinnamon; margin incurved at first, translucent striate, pale orange to cinnamon.

Gills: decurrent, subdistant, narrow, intervenose, light yellow to cream color.

Stalk: ⅝–1⅛″ (1.7–3 cm) long, 1⁄16–⅛″ (0.8–2 mm) wide, equal, enlarging toward the base, dry, nearly glabrous, reddish cinnamon to red-brown, with scattered translucent fibrils, bright yellow-orange fibrils surrounding the base, becoming rusty brown in age; white rhizomorphs at base.

Flesh: thin, firm, cream color; odor not distinctive; taste mild then faintly bitter.

Technical Features: pileipellis short, rectangular, slightly thick-walled hyphae 4–7 µm diameter, orange to reddish brown or dingy yellow-brown, blending into red-brown; **pileotrama** 4–14 µm diameter, red-brown in Melzer's solution, yellow-brown in 3% KOH; **lamellar trama** irregular to loosely interwoven hyphae 2–13 µm diameter with thickened walls, orange-brown in Melzer's solution, yellowish in 3% KOH; **clamp connections** present; **pleurocystidia** 20–27 × 3.5–4.5 µm, fusiform, clavate to cylindric, thin-walled; **cheilocystidia** 20–27 × 4.2–8.4 µm, subventricose, fusiform, clavate, thin-walled; **caulocystidia** 19–38 × 6–13 µm, subventricose, narrowly fusiform, clavate, thin-walled, single or in small fascicles; **basidia** 21–25 × 4–5 µm, clavate, thin-walled, 4-spored, hyaline; **spores** 4.2–6 × 2.5–3.5 µm, elliptical, thin-walled, amyloid.

Spore Print: white.

Fruiting: densely gregarious to caespitose in erect clusters on hardwood, often oak stumps and limbs; fruiting from early summer through late fall; distributed throughout eastern North America; common.

Edibility: unknown.

Observations: Unlike the closely related *X. campanella*, *X. kauffmanii* occurs on hardwood logs in tight erect clusters and not recurved from the wood. In addition, the bright rusty orange cap, smaller spores, lighter pileotrama, and honey-yellow mycelium at the base of the young mushrooms are distinctive characteristics of *X. kauffmanii*. Redhead (1987) reports this species from Costa Rica and "a disjunct Japanese distribution." He also has restudied the Canadian populations and determined that many of the collections should be referred to *X. campanella* and *X. brunneola*.

Xeromphalina tenuipes (Schwein.) A. H. Sm., *Pap. Mich. Acad. Sci. 1*, 38: 84–85. 1953.

Agaricus tenuipes Schwein., *Trans. Amer. Phil. Soc., 2,* 4: 147. 1832.

Cap: 1–2¾″ (2.5–7 cm) broad, obtuse to convex, nearly plane to broadly umbonate in age, dry, velvety, somewhat rugose at times, orange-brown, with an olive-brown sheen; margin becoming translucent-striate in age.

Gills: bluntly adnate with a fine decurrent line, close to subdistant, moderately broad, edges even, white becoming pale yellowish in age.

Stalk: 2–3⅛″ (5–8 cm) long, ⅛–⅜″ (3–8 mm) thick, nearly equal, enlarging somewhat at apex and base, pliant and tough, dry, with a velvety tomentum over entire surface to pubescent at the base, concolorous with the cap; rhizomorphs absent.

Flesh: pliant in cap, watery brown, firm and brown in the stalk; odor none; taste unknown.

Technical Features: **pileipellis** an interwoven layer of pigmented hyphae 2.5–5 µm diameter, thick- and thin-walled, red in 3% KOH with fascicles of **pileocystidia** 29–90 × 2.5–7 µm, narrowly clavate to hypha-like, thin-walled, red to red-brown in 3% KOH; **pileotrama** of interwoven hyphae 1.7–7 µm diameter, wine-red near surface in 3% KOH to yellow below; **lamellar trama** of interwoven hyphae, thin- and thick-walled, 2–7 µm diameter, yellow-brown; **clamp connections** present; **pleurocystida** 23–30 × 5-8 µm, clavate to fusiform, hyaline; **cheilocystidia** 17–37 × 1.5–3.5 µm, narrowly clavate, contorted, branched to irregular, thin-walled, hyaline; **caulocystidia** 21–53 × 3.5–5 µm, nearly cylindric, thick-walled, yellow to red-brown in 3% KOH; **basidia** 20–30 × 6–7 µm, clavate, thin-walled, 4-spored, yellowish in 3% KOH and Melzer's solution; **spores** 6.5–9 × 3.5–5 µm, elliptical, thin-walled, amyloid.

Spore Print: white.

Fruiting: scattered to subcaespitose on hardwood debris, limbs, logs, and stumps; from June to early September; distribution in eastern Canada and the United States; infrequent.

Edibility: unknown.

Observations: A convex-umbonate cap with adnate gills, red-brown pileocystidia and caulocystida, and red pileipellis in 3% KOH are unique features which distinguish *X. tenuipes* from the other species in the genus. In addition, the elliptical spores (3.5–5 µm wide) are wider than any other species of *Xeromphalina* except for the very minute *X. picta* (Fr.) A. H. Sm. Redhead (1987) has provided additional distributional information for the species and also noted that there are no rhizomorphs present.

Aphyllophorales
Cantharellaceae

Gomphus bonarii (Morse) Singer, *Lloydia* 8: 140. 1945.

Cantharellus bonarii Morse, *Mycologia* 22: 219–229. 1930.

Cap: 1⅜–5½″ (3.5–14 cm) wide, slightly depressed at the center when young, becoming moderately depressed at maturity; surface composed of large, tufted, partially erect and typically blunt scales; scales dull orange at the tips and yellow at the bases when fresh, becoming pale orange-brown to dark brown in age or when dry; margin incurved to slightly inrolled when young, somewhat flaring and wavy at maturity.

Ridges: strongly decurrent and often extending half-way down the stalk, consisting of very narrow blunt radial folds with many interconnecting veins, subdistant, milk-white to creamy white, slowly staining pinkish brown when bruised or at maturity.

Stalk: 1⅜–4½″ (3.5–11.5 cm) long, ⅜–1⅜″ (1–3.5 cm) thick, solid, smooth, broad at apex, narrowing toward the base, white with pinkish brown stains; sometimes several fused at the base.

Flesh: moderately thick, firm, tapering toward the cap margin; odor and taste not distinctive.

Technical Features: **pileotrama** interwoven, hyaline, lacking clamp connections; **pleurocystidia** and **cheilocystidia** not differentiated; **basidia** clavate, 2- to 6-spored; **spores** 10–14 × 5–6 µm, subellipsoid, smooth to slightly roughened.

Spore Print: pale ochraceous.

Fruiting: scattered, in groups or in clusters on the ground, often in deep humus under conifers; May-October; Rocky Mountains and Pacific Coast ranges; frequent to infrequent.

Edibility: not recommended. This species has been reported as edible for most people, but cases of gastrointestinal upset have been reported.

Observations: The original description of this species was based on two collections from General Grant National Park, California, which had large clusters of fruiting bodies borne on a common base. *Gomphus floccosus* (Schwein.) Singer is similar but is deeply depressed to funnel-shaped and has yellow to cream or ochre blunt ridges that are decurrent nearly to the base of the stalk; the scales on the surface are thin and often uplifted but are not large, tufted and blunt like those of *G. bonarii*.

Notes: For additional information see Morse (1930) and Smith and Morse (1947).

Ramaria stuntzii Marr, *Bibliotheca Mycologica* 38: 118–120. 1973.

Fruiting Body: 2⅜–6⅝″ (6–17 cm) tall, 1½–5½″ (4–14 cm) wide, consisting of a cauliflower-like mass of compact branches arising from a very massive base.

Stalk: ¾–2¾″ (2–7 cm) tall, 1–2¾″ (2.5–7 cm) wide, solid, cylindrical, very massive, white on the lower portion, with small abortive or primordial white branches, pale orange on the upper portion, developing furrows as branches begin to emerge.

Branches: scarlet, fading to pale orange-red, darkest at the tips; main branches divide to form several smaller branches which also repeatedly subdivide.

Flesh: yellowish white and fibrous in the stalk, rapidly and strongly amyloid when a drop of Melzer's solution is applied to a freshly cut surface; a green reaction occurs on the hymenial surface when a drop of 10 percent aqueous $FeSO_4$ is applied; grayish red to grayish orange and brittle in the branches; odor not distinctive; taste slightly bitter.

Technical Features: **basidia** clavate, 45–75 × 7–12 µm, mostly 4-spored; **spores** 7–10 × 3–5 µm, subcylindric, ornamented with small lobed warts.

Spore Print: orange-yellow.

Fruiting: solitary or in groups on the ground under western hemlock; September-November; Pacific Northwest from British Columbia south to California; uncommon.

Edibility: unknown.

Observations: A truly beautiful species which often produces a large number of fruiting bodies in a fairy ring. This species is best recognized in the field by noting the white stalk base, the pale orange upper stalk and the scarlet branch tips. *Ramaria cyaneigranosa* Marr and Stuntz and *R. araiospora* Marr and Stuntz are similarly colored but are less robust and lack the strongly amyloid reaction on the fresh flesh of the stalk.

Notes: Additional information is provided by Marr and Stuntz (1973).

Aphyllophorales
Hydnaceae

Hydnellum caeruleum (Hornem. ex Pers.) P. Karst., *Medd. Soc. Fauna Fl. Fenn* 5: 41. 1879.

Hydnum caeruleum Hornem., *Flora Danica VIII, Fasc. 23*, pl. 1374. 1808.

Cap: $1\frac{3}{16}$–$4\frac{3}{8}$″ (3–11 cm) wide, convex, plane in age, dry, soft and velvety to touch, glabrous, solitary or somewhat confluent, not zonate, whitish, soon tinted blue, often becoming almost entirely blue, gradually changing to brown or dark brown in the center.

Spines: $\frac{1}{8}$–$\frac{3}{8}$″ (3–8 mm) long, crowded, slender, decurrent, pale fawn becoming dark brown with light tips.

Stalk: $\frac{3}{4}$–$2\frac{3}{8}$″ (2–6 cm) long, $\frac{5}{16}$–$\frac{3}{4}$″ (7–20 mm) wide, always distinct (never fused), enlarging downward and typically bulbous or tapering downward, irregular or somewhat flattened, dry, glabrous, pale buff at first, becoming darker brown in age.

Flesh: in cap with a soft upper layer and a tough, compact lower layer, zonate, blue to purple with brown; in stalk homogeneous or somewhat zoned, orange to orange-red, sometimes with blue, tough; odor none or slightly of cooked meat; taste mild or slightly acid.

Technical Features: **pileipellis** of interwoven hyphae 2.5–6 μm diameter, thin- and thick-walled; **pileotrama** more densely interwoven with more thick-walled hyphae; **clamp connections** scattered in the pileipellis; **cystidia** absent; **basidia** 27–36 × 3.5–5.5 μm, clavate, 4-spored, no basal clamp connections; **spores** 4.5–6 × 3.5–5.5 μm, subglobose to oblong, tuberculate-warted, light brown in 3% KOH.

Spore Print: light brown.

Fruiting: usually gregarious, rarely single, caps often somewhat fused; on ground under pines, spruce, and other conifers; August to October; distributed in southeastern to northeastern North America across Canada and the northern United States to the Rocky Mountains and western coastal conifer forests; frequent.

Edibility: inedible.

Observations: The young caps tinted blue to dark blue, then buff to brown in age and the absence of a fragrant odor are distinctive characters which distinguish *H. caeruleum* from other species of *Hydnellum*. *Hydnellum suaveolens* (Scop. ex Fr.) Karst. has a deep violet stalk and a very fragrant odor (Harrison 1961).

Hydnellum regium Harrison, *Can. J. Bot.* 42: 1231–1233. 1964.

Fruiting Body: 6–10″ (15–25 cm) wide, 3½–6″ (9–15 cm) high, very complex, composed of numerous overlapping caps forming rosettes; caps covered with matted hairs, sometimes eroded, spongy, violaceous-black, with concentric ridges and obscure zonations, becoming radially striate toward the margin; margin rounded, grayish white becoming brownish gray to dark olive-brown in age.

Spines: ⅛–¼″ (3–6 mm) long, decurrent, very close, pale violet to dark purple when young, becoming grayish brown in age.

Stalk: central, complex, giving rise to and continuous with the overlapping caps, stout, solid, tapering to a narrow cord below, pinkish cinnamon, sometimes with small cavities containing orange-yellow mycelium that are evident when the stalk is cut lengthwise.

Flesh: firm, brittle, color variable from pale brown to violaceous-black, zonate; odor variable from pungent aromatic for some specimens to faint or absent in others; taste variable from disagreeable to slightly bitter or not distinctive.

Technical Features: **pileotrama** of interwoven hyphae up to 5.5 µm diameter, uninflated, with clamp connections; **basidia** 25–33 × 5.2–6.4 µm, clavate, 4-spored; **spores** 4.5–6 × 3.5–4.5 µm, oblong to subglobose, tuberculate; tubercles low, rounded.

Spore Print: pale brown.

Fruiting: solitary or in groups on duff under conifers; September-November; Pacific Northwest from British Columbia south to Oregon, Idaho and Colorado; common on the Pacific Coast but uncommon in the Rocky Mountains.

Edibility: inedible.

Observations: A very large and beautiful species. The combination of the large size, complex arrangement, dark caps with paler colors in the stalk and brittle flesh is distinctive.

Notes: This species was first described by Harrison (1964).

Phellodon atratus K. A. Harrison, *Can. J. Bot.* 42:1209. 1964.

Cap: ¾–1½″ (2–4 cm) broad, single, several or several fused, plane or somewhat depressed to umbilicate, smooth or with minute hairs over the center, zonate, purple to purple-gray, blackish purple when handled; margin light bluish gray, uneven and undulating; drying brownish drab to vinaceous-drab.

Spines: short decurrent, up to 2 mm long, tapering to a point, lavender to lavender-blue, bruising blackish brown.

Stalk: 1½–2″ (4–5 cm) long, ⅛–¼″ (3–7 mm) wide, tapering downward, dry, smooth or minutely roughened, purplish gray to black, bruising blackish brown.

Flesh: duplex in cap with a thin spongy upper layer and a fibrous lower layer, blue-black; stalk also duplex, purplish to black; odor of sweet clover or burnt sugar fresh and dried; taste mild.

Technical Features: pileipellis a loosely interwoven trichodermium of hyaline hyphae, 3.6–6.3 μm diameter, light brown with encrusted material in 3% KOH; **pileotrama** of interwoven hyphae 3.5–6.5 μm diameter, brownish, thin-walled and scattered thick-walled cells; **clamp connections** absent; **basidia** 23–38 × 4–7 μm, clavate, thin-walled, 4-spored; **spores** 3.3–4.5 × 3.3–3.8 µm, finely warted, globose to subglobose, light brown in 3% KOH.

Spore Print: white.

Fruiting: usually gregarious, often caps or stalks fused in twos or threes, on ground under conifers, often second growth; in early to late fall; distributed in the coastal forests of the Pacific Northwest; common.

Edibility: inedible.

Observations: *Phellodon atratus* has a combination of a purple to purple-gray cap with a bluish-gray margin, lavender to lavender-blue spines and a cap diameter of ¾–1½″ (2–4 cm). The closely related *P. melaleucus* (Fr.) Karst. has a dark vinaceous-brown sometimes tinted violaceous-blue cap, a whitish margin, spines which are ash-gray or violet-white to white and a smaller cap, seldom over 1⅛″ (3 cm) wide (Hall and Stuntz 1971). Both species have similar distribution ranges and both species have the typical sweet clover smell and become darker when bruised.

Lentinellus montanus O.K. Miller, *Mycologia* 57: 933–945. 1965.

Cap: 1¾–2⅝″ (4.5–6.5 cm) broad, 1½–4¼″ (4–11 cm) wide, sessile, shell-shaped to somewhat fan-shaped, moist but not viscid, hirsute to villose over the center to glabrous elsewhere, dark brown to red-brown; margin glabrous or nearly so, light pinkish cinnamon to pale pinkish buff, often crenulate in age.

Gills: broad, subdistant, alternate with long lamellulae, edges coarsely serrate, white, tinted purplish at first to buff in age.

Flesh: tough, hygrophanous, brown to light brown, 1⁄16–3⁄16″ (1–4 mm) thick; odor none or slightly aromatic; taste mild or somewhat acrid.

Technical Features: pileipellis in center a loose trichodermium arising from decumbent hyphae 3–9.2 µm diameter, yellowish in 3% KOH and Melzer's solution; **pileotrama** of interwoven thick-walled hyphae 3–7.7 µm diameter, yellowish in Melzer's solution; **lamellar trama** subparallel to irregular in center, 4.5–8 µm diameter, yellowish in Melzer's solution; **pleurocystidia** and **cheilocystidia** 26–39 × 4.5–8.4 µm, clavate, fusiform to lageniform with elongated, mucronate apices on many, thin-walled, protruding beyond hymenium, hyaline, sparse to abundant;

basidia 20–47 × 5.2–8 μm, clavate, thin-walled, 4-spored, hyaline; **spores** 4.5–6.5 × 4–5 μm, ovoid to nearly globose, thin-walled, hyaline with blunt echinulations, amyloid in Melzer's solution.

Spore Print: cream to buff.

Fruiting: solitary but most commonly in imbricate clusters on conifer logs, stumps, limbs, and twigs, under and near melting snow banks from 5-10 thousand feet elevation; May to July; distributed in the northern and central Rocky Mountains as far south as Utah and west to the Cascade Range (known from Washington, Idaho, Montana, Colorado, Utah, Wyoming, and California); frequent to common.

Edibility: inedible.

Observations: The unusual habit and habitat, large spores and pleurocystidia, and the lack of amyloid tramal tissue are a combination of characters which separates *L. montanus* from other species of *Lentinellus* (Miller 1965a; Miller and Stewart 1971).

Notes: The Agaricales have thin-walled tramal hyphae. Species of *Lentinellus* are therefore placed in the family Lentinellaceae Kotl. and Pouz. (Jülich 1981).

Pleurotus cystidiosus O. K. Miller, *Mycologia* 61: 889–893. 1969.

Cap: ¾–2⅜″ (2–6 cm) broad, broadly convex, plano-convex in age, dry, with numerous, appressed squamules, creamy white, light brown to brown; margin incurved at first, often slightly crenulate in age.

Gills: decurrent, subdistant, ventricose, with two tiers of irregular lamellulae, white to buff at maturity.

Stalk: ¼–⅝″ (6–15 mm) long, 3⁄16–¾″ (5–20 mm) wide, tapering to a narrow base, eccentric, dry, shiny white covered with a minute white pubescence; base with copious, white mycelium and rhizomorphs.

Flesh: firm and white throughout; odor not distinctive; taste pleasant.

Technical Features: pileipellis a trichodermium of numerous **pileocystidia** 28–42 × 8.5–12 µm, clavate to cylindric, thin-walled, in loosely formed fascicles; **pileotrama** of interwoven hyphae 2.5–12 µm diameter, thin-walled, hyaline; **lamellar trama** of cellular elements 2.5-7 µm diameter, thick- and thin-walled, angular; **clamp connections** present; **cheilocystidia** and **pleurocystidia** 23–57 × 8.5–17 µm, pyriform, clavate to fusiform, thin-walled, hyaline; **basidia** 35–50 × 6.5–9.5 µm, clavate, thin-walled, 4-spored; **spores** 11–17 × 4.2–5 µm, subfusiform in profile, oblong-elliptical in face view, thin-walled, yellowish in Melzer's solution.

Spore Print: white.

Fruiting: single to several on a wide variety of living hardwood hosts including red maple, red gum, black oak, cottonwood, and willow; usually found in late summer and fall; distributed throughout the hardwood forests of eastern North America especially the southern and southeastern United States; infrequent.

Edibility: edible and choice.

Observations: The abundant, clavate pileocystidia, oblong-elliptical inamyloid spores, and short pyriform cheilocystidia are distinctive characters of *P. cystidiosus*. The presence in culture of minute coremia with round black tips around the base of many fruiting bodies is a unique, asexual characteristic of *P. cystidiosus* (Miller 1969; Pollack and Miller 1976). The asexual stage, known as *Antromycopsis broussonetiae* Pat. and Trab., is very distinctive among basidiomycete anamorphs. *Pleurotus abalonus* Han, Chen and Cheng has also been described from Asia but may be conspecific with *P. cystidiosus*. It is raised commercially for food in Asia, especially in Thailand, Taiwan and China (Stamets 1993).

Notes: Thick-walled tramal hyphae in the lower cap and stalk is not a feature of the Agaricales. Species of *Pleurotus* are now placed in the Polyporaceae Kühner and Romag. ex Kühner (Jülich 1981).

Pleurotus populinus Hilber and O. K. Miller, *Can. J. Bot.* 71: 127. 1992.

Cap: 1½–7⅛″ (4–18 cm) broad, 1½–5⅛″ (4–13 cm) wide, convex, oyster shell to fan-shaped, smooth, pubescent over the lateral point of attachment, dry, ivory-white, pinkish buff, pale orange to orange-gray; margin incurved at first, crenulate in age.

Gills: decurrent over the short point of attachment, subdistant, broad, ¼–⅜″ (5–10 mm), ventricose, with two tiers of lamellulae, white to cream color, tinted pink between the gills.

Stalk: lateral point of attachment, ⅜–1″ (1–2.5 cm) long, 3/16–15/16″ (6–24 mm) wide, tapering from cap to host, dry, pubescent, white, ridged from terminus of the gills.

Flesh: soft, firm, white in cap and stalk; odor pleasant, fungoid to anise-like; taste pleasant.

Technical Features: pileipellis of a dense, tightly interwoven layer of yellowish hyphae 3–7 μm diameter, thin-walled; **pileotrama** interwoven, thick- and thin-walled, hyaline hyphae 3–8 μm diameter; **lamellar trama** nearly parallel, thick- and thin-walled, hyaline hyphae 3–6 μm diameter; **clamp connections** present on all tissues; **cheilocystidia** and **pleurocystidia** 19–38 × 4–7 μm, thin-walled, hyaline, fusiform to ventricose-rostrate, or capitate to tenpin shaped; **basidia** 20–27 × 5–6 μm, clavate, 4-spored, hyaline; **spores** 9–15 × 3–5 μm, narrowly elliptical to oblong, thin-walled, hyaline in 3% KOH and Melzer's solution.

Spore Print: buff.

Fruiting: single to numerous, often imbricate on limbs, stumps, or logs of hardwood trees, especially aspen and black cottonwood; in June and July; distributed in most of the northern United States and Canada and in the montane and northern regions of western North America; frequent.

Edibility: edible and choice.

Observations: The large spores, northern distribution on aspen and cottonwood, and the buff spore print are the chief characters which distinguish *P. populinus* from the closely related but small spored *P. ostreatus* (Jacq. ex Fr.) P. Kumm. and *P. pulmonarius* (Fr.) Quél. Genetic testing by Vilgalys et al. (1992) has shown that there are three intersterile species in North America with a tendency for *P. ostreatus* to be found in eastern North America. *Pleurotus pulmonarius* is more generally distributed in North America and is found on a variety of hosts in contrast to *P. populinus*.

Notes: Thick-walled tramal hyphae in the lower cap and stalk is not a feature of the Agaricales. Species of *Pleurotus* are now placed in the Polyporaceae Kühner and Romag. ex Kühner (Jülich 1981).

Albatrellus dispansus (Lloyd) Canf. and Gilbn., *Mycologia* 63: 965. 1971.

Polyporus dispansus Lloyd, *Mycol. Writings 3, Stipitate Polypores*, p. 192, 4. 498. 1912.

Fruiting Body: multipileate up to 14″ (35 cm) wide, depressed, oval, with individual caps ¾–2⅜″ (2–6 cm) wide, densely imbricate, dry, finely pubescent to tomentose, azonate, yellow at first, tinted orange in age.

Stalk: broad, central, short, up to 2¾″(7 cm) wide, white, smooth.

Pores: 2–3 per mm, angular; surface white, minutely fimbriate, not bruising.

Flesh: firm, dry, soft and pliant to brittle when fresh but very hard when dried; odor fragrant, like pears; taste mild.

Technical Features: **pileipellis** of puzzle-like cells 6.5–20 μm diameter, slightly thick-walled, hyaline in 3% KOH; **pileotrama** similar but light reddish in 3% KOH; **pore trama** densely interwoven, mostly filamentous; **clamp connections** absent; **cystidia** absent; **basidia** 17–20 × 5–6 μm, clavate, thin-walled, usually 2-spored, some 4-spored; **spores** 4–5 × 3–4 μm, ovoid, hyaline, thin-walled, inamyloid.

Spore Print: white.

Fruiting: single, on ground under Rocky Mountain maple and various conifers including ponderosa pine, Douglas fir, cedar, southwestern white pine, often in old growth stands in the fall; known from Idaho and Arizona; rare.

Edibility: inedible.

Observations: The imbricate fruiting body with numerous yellow caps, small pores, a white pore surface, small ovoid inamyloid spores, and the limited distribution in Idaho and Arizona are distinctive features of this polypore. Gilbertson and Ryvarden (1986) report it from Arizona and Tylutki (1987) has an account of it from Idaho. Canfield and Gilbertson (1971) and Canfield (1981) report that it causes a brown cubical rot of conifers and decays Douglas fir test blocks.

Note: This species was described by Overholts (1953) as *Polyporus illudens* Overh. (an invalid name).

Albatrellus ovinus (Schaeff.: Fr.) Murr., *J. Mycol.* 9: 91. 1903.
Polyporus ovinus (Schaeff.) Fr., *Syst. Mycol.* 1: 346. 1821.

Cap: 2–6″ (5–15.5 cm) wide, convex and circular to kidney-shaped or fused and irregular, sometimes slightly depressed; surface dry, whitish to creamy yellow or pale grayish, smooth when young becoming areolate with pale yellow flesh showing in the cracks in age; margin incurved and entire, becoming uplifted, wavy and partially eroded at maturity.

Pores: small, 2–5 per mm, angular near the stalk, round near the margin; surface white to pale yellow, not staining when injured; tubes decurrent, 1⁄16–3⁄16″ (1.5–4.5 mm) deep, whitish.

Stalk: 1–3″ (2.5–7.5 cm) long, 3⁄8–1 3⁄8″ (1–3.5 cm) thick, enlarging downward or nearly equal, typically ventricose at the base, central or eccentric; surface white to pale brownish yellow, smooth or very finely tomentose.

Flesh: white, thick, firm, unchanging when exposed, slowly drying yellowish or greenish yellow; odor not distinctive; taste mild.

Technical Features: **trama** monomitic, thin-walled, simple-septate; **cystidia** absent; **basidia** clavate, 4-spored; **spores** 3–5 × 3–3.5 µm, ovoid to subglobose, smooth, thin-walled, hyaline, inamyloid.

Spore Print: white.

Fruiting: solitary, scattered or several fused together on the ground near conifers; July-December; widely distributed in North America: eastern Canada to Tennessee west to British Columbia south to California, also the mountains of Utah, Colorado, Arizona and New Mexico; infrequent to common.

Edibility: edible.

Observations: The whitish, often circular cap, white pore surface, small pores, inamyloid spores, simple septa, white central stalk and growth under conifers distinguish this species. *Albatrellus confluens* (Albertini and Schwein.: Fr.) Kotl. and Pouz. is similar but forms orange to pinkish buff fruiting bodies which are typically clustered, with a bitter to cabbage-like taste. *Albatrellus subrubescens* (Murr.) Pouz. has amyloid spores and clamp connections and is similar in other ways but rare.

Notes: This species is commonly called the "Sheep Polypore." For more information see Gilbertson and Ryvarden (1986).

Neolentinus ponderosus (O. K. Miller) Redhead and Ginns, *Trans. Mycol. Soc. Japan* 26: 357. 1985.

Lentinus ponderosus O. K. Miller, *Mycologia* 57: 941–943. 1965.

Cap: 2–14½″ (5–33 cm) broad, broadly convex to nearly plane in age, dry, with appressed, cinnamon-brown squamules, flesh in between pinkish buff; margin often slightly incurved at first.

Gills: close, adnate, narrow, edges coarsely serrate, alternating with lamellulae which extend up to two-thirds toward stalk, bright to pale white, light buff to light orange in age.

Stalk: 1–4″ (2.5–10 cm) long, 1¼–4″ (3–10 cm) wide, dry, finely squamulose and white to buff at apex, forming small appressed reddish brown squamules which coalesce over the lower half forming a reddish brown surface; partial veil absent and no annulus present.

Flesh: firm to tough in age, white to buff throughout, unstained in Melzer's solution; odor fruity or not distinctive; taste mild.

Technical Features: **pileipellis** of innate, fibrillose hyphal cells 4.4–8 μm diameter; **pileotrama** of interwoven, thick- and thin-walled inamyloid hyphae 2.5–6 μm diameter, **clamp connections** numerous; **cheilocystidia** and **pleurocystidia** narrowly clavate to hypha-like, 20–64 × 3.5–6.0 μm, thin-walled, hyaline; **basidia** 26–36 × 5–8.8 μm, clavate, thin-walled, 4-spored; **spores** 8–10.5 × 3.5–4.4 μm, thin-walled, subfusiform in profile, elliptical in face view, inamyloid.

Spore Print: dull white to buff.

Fruiting: single to several on conifer logs and stumps, especially ponderosa pine, most often in openings or cut over areas; July through August; distributed throughout Idaho, Montana, Oregon, Washington, and California; infrequent to frequent.

Edibility: edible when young and tender.

Observations: This is a large, robust species closely related to *Neolentinus lepidius* (Fr.: Fr.) Redhead and Ginns from which it differs by the lack of a partial veil, larger size with a thick stalk, lack of thickened recurved squamules on the cap and stalk, and the narrow, close lamellae. In an unusual fruiting, specimens of *N. ponderosus* found on very large logs in the Idaho Primitive Area had caps which ranged from 11–16½″ (29–42 cm) in diameter.

Notes: *Neolentinus* was proposed by Redhead and Ginns (1985a) for those species which caused brown rots in wood. The thick-walled tramal hyphae throughout the cap and stalk is not a characteristic of the Agaricales. This species is a gilled polypore according to Jülich (1981).

Aphyllophorales
Polyporaceae

Nigroporus vinosus (Berk.) Murr., *Bull. Torrey Bot. Club* 32: 361. 1905.

Polyporus vinosus Berk., *Ann. Mag. Nat. Hist., Ser.* 2, 11: 195. 1852.

Fruiting Body: pileate, effused-reflexed, semicircular, often depressed, ¾–4¾″ (2–12 cm) broad; surface dry, leathery, felty, radially appressed, fibrillose, with narrow zones or azonate, red-brown, purplish brown, dark violet, black-brown; with a central or eccentric attachment to the substrate.

Pores: minute 7–8 per mm; tubes up to ⅛″ (3 mm) long; surface purple-brown, dark violet, dark brown, thickest at base, thin at margin.

Flesh: rigid, tough, somewhat pliant but brittle when dry, dark brown to red-brown; odor not distinctive; taste unknown.

Technical Features: pileipellus and **pileotrama** of thin-walled hyphae 2–4 µm diameter, with thick-walled skeletal hyphae 2–6 µm diameter, pale pinkish brown in 3% KOH; **clamp connections** present; **cystidia** absent; **basidia** 6–10 × 3–4 µm, clavate, thin-walled, 4-spored; **spores** 3–4.5 × 1–1.5 µm, allantoid, thin-walled, hyaline, inamyloid.

Spore Print: white.

Fruiting: single or more commonly several together on dead limbs, logs, and stumps of hardwoods, rarely conifer wood (noted on oak, sweet gum and pine); in late summer and fall; distributed in the southeastern United States from Virginia south to Florida and west to Louisiana and Texas; infrequent.

Edibility: inedible.

Observations: The dark brown, purple-brown to red-brown cap and pore surface, combined with the minute pores (7–8 per mm), and small, allantoid spores are a combination of characters which distinguish *Nigroporus vinosus*. This unusual species is a white rotting, decay fungus which is also distributed throughout the tropics (Gilbertson and Ryvarden, 1987, 454–455) and the only one in the genus in North America.

Oligoporus leucospongia (Cke. and Harkn.) Gilbn. and Ryv., *Mycotaxon* 22: 365. 1985.

Polyporus leucospongia Cke. and Harkn., *Grevillea* 11: 106. 1883.

Spongiporus leucospongia (Cke. and Harkn.) Murr., *Bull. Torrey Bot. Club* 32: 474. 1905.

Fruiting Body: sessile, usually solitary, elongated, 1⅛–4″ (3–10 cm) long, ⅜–1½″ (1–4 cm) wide, ¼–1″ (5–25 mm) thick, somewhat rounded, soft and watery when fresh, surface smooth to slightly rugose, white sometimes tinted dull cinnamon-pink; margin extends down partially over the pore surface and is sometimes reddish brown.

Pores: 3–4 per mm, circular to angular, minutely roughened; surface soft, white to cream color.

Flesh: white, azonate, up to ⅝″ (1.5 cm) thick, two-layered, with a soft, cottony, upper layer and a firm zone just above the pores; odor not distinctive; taste mild.

Technical Features: pileipellis a tightly interwoven layer of dingy yellow hyphae 4.5–6 μm diameter, thin- to thick-walled; **pileotrama** of loosely interwoven, hyaline, thin- to thick-walled hyphae 3–7 µm diameter; **clamp connections** present on all tissues; **cystidia** absent; **hyphal pegs** present, narrow, pointed, composed of clavate, thick-walled hyphae 15–30 × 3–7 µm; **basidia** 16–23 × 4–5 µm, clavate, thin-walled, 4-spored, with basal clamps; **spores** 4.5–6 × 1–1.4 µm, allantoid, hyaline, inamyloid.

Spore Print: white.

Fruiting: single to several on decorticated conifer logs, stumps, and limbs; Engelmann spruce and subalpine fir are common hosts but on other conifers as well; fruiting on wood covered with snow, reaching maturity shortly after the snow recedes; from May to July; distributed from northern New Mexico and Arizona north to southern Alberta and British Columbia, and the Northwest Territory in Canada; common.

Edibility: inedible.

Observations: The white, cottony, soft cap, and sessile habit on decorticated conifer logs in the high elevation forests of the western United States and adjacent Canada is the distinctive combination of characters of *O. leucospongia*. Successful fruiting occurs when snow is sufficiently deep at high elevations so that daily melting occurs slowly to stimulate fruiting under cool moist conditions. Complete fruiting bodies can be produced with temperatures less than 7°C (unpublished data). Hyphal growth is often maximized at about 12–16°C which places this fungus in the group of cold-loving or psychrophilic fungi. We have no records of this fungus from Alaska, and it is not listed by Volk et al. (1994).

Notes: The illustration (546 *Spongiporus leucospongia*) in Lincoff (1981) is red-brown and probably *Oligoporus fragilis* (Fr.) Gilbn. and Ryv. but not *O. leucospongia*. Both of the species were placed in *Spongiporus* by Murrill. Gilbertson and Ryvarden (1987, 456–491) have keys and descriptions of the North American species of *Oligoporus*.

Aphyllophorales
Polyporaceae

Polyporus tenuiculus (P. Beauv.) Fr., *Syst. Mycol.* 1: 344. 1821.
Favolus brasiliensis (Fr.) Fr., *Elench. Fung.* 1: 44. 1828.
Hexagonia reniforme Murr., *North Am. Fl.* 9: 50. 1907.

Cap: ¾–4½″ (2–11.5 cm) wide, kidney-shaped to fan-shaped or shell-shaped, broadly convex to nearly plane, soft and flexible when fresh; surface dry, smooth to finely tomentose, often with radial lines over the disc and extending down onto the stalk, white to pale grayish white; margin entire when young, becoming lobate and wavy in age, thin, with tiny projecting hairs.

Pores: 1/16–⅛″ (1.5–3 mm) long, radially elongated, sometimes nearly hexagonal, largest over the stalk; tubes strongly decurrent often to the base of the stalk, 1/16–⅛″ (1–3 mm) deep, whitish; pore surface white to pale grayish white, not staining when injured.

Stalk: ¼–1″ (6–25 mm) long, ⅛–¼″ (3–6 mm) thick, tapering downward, lateral to eccentric, sometimes indistinct; surface finely tomentose to fibrillose, white, with a white basal mycelium.

Flesh: whitish in cap and stalk, thin, soft, unchanging when cut or bruised; odor and taste not distinctive.

Technical Features: hyphal system dimitic; generative hyphae hyaline, thin-walled and with clamps; vegetative hyphae resembling skeletal hyphae and intermediate between skeletal and binding hyphae; **cystidia** absent; **basidia** clavate, 20–30 × 4–7 μm, with oil-drops, 4-spored; **spores** 9–12 × 2–4 μm, cylindrical to subelliptical, smooth, thin-walled, hyaline, typically with 2–3 oil drops.

Spore Print: whitish.

Fruiting: solitary, imbricate or in clusters on decaying hardwood; throughout the year; common from Florida to Texas and very common throughout the tropics.

Edibility: inedible.

Observations: The whitish, kidney-shaped to fan-shaped cap and elongated to nearly hexagonal pores are distinctive characters. It causes a white rot in dead hardwoods. Several species of hardwoods are substrates for this polypore.

Notes: For more information see Gilbertson and Ryvarden (1987, 667–668).

Hymenogastrales
Rhizopogonaceae

Truncocolumella citrina Zeller, *Mycologia* 31: 1–32. 1939.

Fruiting Body: ¾–2¾″ (2–7 cm) broad, ⅝–1¾″ (1.6–4.5 cm) tall, obovate to irregular, surface smooth, glossy or dull, dry, becoming somewhat wrinkled or areolate in age, whitish to pale yellow at first, then pale golden yellow with orange-yellow streaks at maturity, slowly staining dull orange where bruised.

Stipe Columella: thick, stump-like or tree-like with branches extending throughout the spore mass, yellowish white to dull yellow.

Sterile Base: formed by the broad base of the stipe-columella, yellow, attached to yellow rhizomorphs.

Spore Mass: pale yellow when young, becoming grayish olive to brownish olive, with small irregular chambers, gelatinous to firm then deliquescent; odor not distinctive; taste mild to somewhat unpleasant.

Technical Features: **peridium** 75–100 μm thick, hyphae interwoven; **basidia** cylindrical, 2- or 4-spored; **spores** 6–10 × 3.5–5 μm, elliptical, smooth, thin-walled, with one oil drop, lacking an apical pore.

Fruiting: on the surface of the ground or partially buried, solitary, scattered, or in clusters in duff under conifers; Pacific Northwest south to California and Idaho; August to November; common.

Edibility: edible but of poor quality.

Observations: The smooth yellow surface, a distinctive stipe-columella, a yellow sterile base attached to yellow rhizomorphs and the somewhat unpleasant taste distinguish this species. It is usually partially buried and one of the most common fungi of the Pacific Northwest. It is mycorrhizal with western conifers.

Notes: For more information see Zeller (1939).

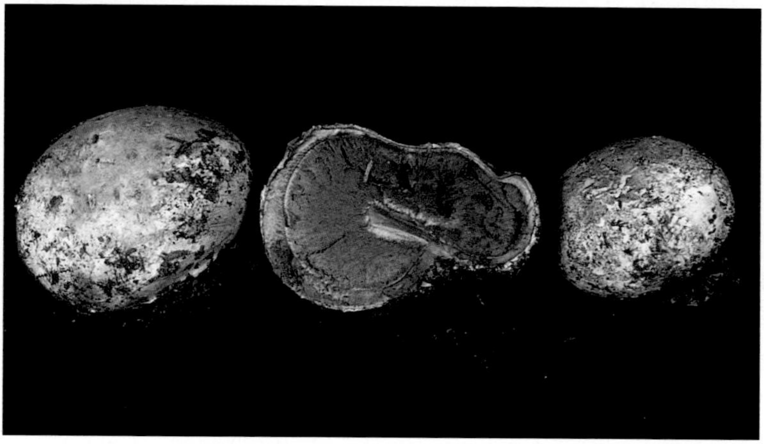

Radiigera fuscogleba Zeller, *Mycologia* 36: 633. 1944.

Fruiting Body: depressed globose, 1½–3″ (4–7.5 cm) broad, 1⅛–2″ (3–5 cm) high, outer skin white, indehiscent, staining pinkish brown and darkening to red-brown when bruised or cut, drying darker brown; skin 3-layered, outer layer about 1 mm thick, middle layer up to 6 mm thick, and inner layer very thin and adhering to the spore mass.

Sterile Base: a central pseudocolumella, firm, white, oval and extending up into the spore mass.

Spore Mass: soft, moist, white at first, becoming light brown as the spores mature, then dark brown in age, oriented in radial rows from the pseudocolumella; odor and taste unknown.

Technical Features: **exoperidium** composed of long, branched, hyaline hyphae 0.5–6 µm diameter; **mesoperidium** of large, irregularly shaped, hyaline cells up to 24 µm diameter; **endoperidium** of long, branched hyphae 0.5–5.5 µm diameter, hyaline with clamp connections; **pseudocolumella** of distinctively branched, hyaline, clavate cells up to 17 µm diameter, with a clamp connection at each septum; **clamp connections** present on all tissues; **basidia** 11–18 × 6.5–8.5 µm, capitate with a swollen base, thin-walled, 4- to 8-spored; **spores** 4–5.5 μm diameter, globose, with truncate-warts, 0.5–1 μm long, yellow-brown to brown when mature.

Fruiting: gregarious, partially hypogeous, emerging near maturity, in rich organic soil, often in decomposing wood, under older stands of Engelmann spruce, Douglas fir, and grand fir in early to midsummer; distribution in Montana, Idaho, Washington, Oregon, and California but most likely throughout the western United States and Canada; rare.

Edibility: inedible.

Observations: The close resemblance to an earth star (*Geastrum*) is documented by Askew and Miller (1977) which includes the distinctive pseudocolumella, capitate basidia, and distinctive warted spores. Unlike the genus *Geastrum*, *Radiigera* does not open up nor does it possess an ostiole (pore) and peristome as illustrated by Miller and Miller (1988). It was transferred by Askew and Miller (1977) to the earthstars (Geastraceae) in the Lycoperdales. The other common species is *Radiigera atrogleba* Zeller which differs by having a jet black spore mass which is easily distinguished from the brown spore mass of *R. fuscogleba*. The spore mass of both species is eaten by rodents and slugs.

Notes: Zeller (1944) placed *Radiigera* in the family Mesophelliaceae of the Lycoperdales on the basis of the lack of dehiscence and an apical pore. No morphological investigation was carried out at that time. Castellano et al. (1989) incorrectly assigned *Radiigera* to the Astraceae in the Sclerodermatales where it has no morphological affinities.

Bovistella radicata (Mont.) Pat., *Bull. de la Soc. Mycol. de Fr.* 15: 54–59. 1899.

Bovistella ohiensis Ellis and Morg. *Jour. Cin. Soc. Nat. Hist.* 14: 141–142. 1892.

Fruiting Body: 1–3½″ (2.5–9 cm) wide, subglobose to broadly pyriform, attached to the soil by a rooting base; entire surface of the exoperidium, or at least the upper half, composed of soft, white pyramidal warts which are often fused at their tips to form fascicles, with granular to scurfy white material distributed nearly overall, becoming minutely scurfy to nearly smooth at maturity, white at first, dull orange-yellow in age, wearing away irregularly to expose the endoperidium; endoperidium thin, papery, splitting open at maturity by an apical pore or slit which eventually enlarges to expose most of the gleba.

Sterile Base: whitish, becoming orange-yellow in age, well developed, cup-like, occupying most of the narrowed, lower portion of the fruiting body.

Spore Mass: white and spongy when young, becoming yellowish to olive, and finally yellow-brown and powdery in age; odor not distinctive; taste of immature flesh mild to slightly sweet.

Rooting Base: ¾–1¾″ (2–4.5 cm) long, ¼–⅝″ (7–16 mm) thick, tapering downward, sometimes forking, whitish, densely fibrillose, typically coated with soil and debris.

Technical Features: capillitium of separate units, highly branched and entangled; main axis 7–12 μm thick; branches tapering to about 3 μm; **basidia** clavate, short, 4-spored; **spores** 3.5–5.5 × 3.5–4.5 μm, globose to subglobose, smooth, hyaline, with one large oil drop and a hyaline, persistent pedicel 6–12 μm long.

Fruiting: solitary, scattered, or in groups on soil in open areas, cultivated fields, pastures and woodlands, especially oak-pine; June-December; New York to Florida, west to Michigan and Texas; infrequent but most commonly encountered in its southern range.

Edibility: unknown.

Observations: The genus *Bovista* lacks a distinctive, well-developed subgleba which is always present in *Bovistella*. Young and fresh specimens of *B. radicata* have a white gleba, lack a distinctive odor and have a mild to slightly sweet taste. *Bovistella echinata* (Pat.) Lloyd also has a white fruiting body but it is floccose, much smaller, ¼–⅜″ (6–9 mm), and is attached to soil by a short pad of fibrils and not a rooting base.

Notes: For more information see Coker and Couch (1928) and Miller and Miller (1988).

Lycoperdales
Lycoperdaceae

Calvatia rubroflava (Cragin) Morgan, *J. Cinc. Soc. Nat. Hist.* 12: 171. 1890.

Lycoperdon rubro-flavum Cragin, *Washburn Lab. Nat. Hist. Bull.* 1: 36. 1885.

Calvatia aurea Lloyd, *Myc. Writings* 1: 11. 1899.

Fruiting Body: subglobose to pyriform, often flattened somewhat at the apex, 1⅛–4¾″ (3–12 cm) broad, ¾–3⅛″ (2–8 cm) tall, smooth to minutely areolate, nearly white tinted pink or lavender, becoming yellow to bright yellow or orange when bruised, cut or rubbed; skin two-layered 1/16″ (1–2 mm) thick, with a very thin outer layer which soon sloughs off.

Sterile Base: sulcate to plicate, usually tapering abruptly to a narrow base with white rhizomorphs, surface staining as described above; flesh yellow to orange-yellow, homogeneous.

Spore Mass: pure white with minute cavities when young, becoming bright yellow-orange to dull orange in age, maturing from the center outward; odor none at first, becoming strong like "old ham" in age, taste mild and pleasant.

Technical Features: peridium 0.2–1.5 mm thick, a polycystoderm of oval to irregular cells 8–11 × 6.3–8.1 µm, thin-walled, hyaline; **capillitium** 2–4 µm diameter, thick-walled, often flexuous, occasionally septate and branched, olive-brown in 3% KOH; **basidia** 10–17 × 7.2–8 µm, 2- to 4-spored, thin-walled, hyaline; **spores** 3–4 µm, globose, minutely warted, nonpedicellate, dark olive-brown in 3% KOH.

Fruiting: several to gregarious in cultivated soil; from late August through the fall; from Massachusetts south to Florida and west to Missouri and Iowa; infrequent but widespread.

Edibility: edible when young and pure white.

Observations: The brilliant yellow to orange stains on the surface, and the yellow-orange maturing sterile base and spore mass are diagnostic features of this unique *Calvatia*. The species is also known from South America, Asia, and Australia. Coker and Couch (1928) have speculated that it may have been introduced to North America from abroad.

Calvatia rubroflava (young specimens)

Calvatia rubroflava (mature specimens)

Notes: This species has also been placed in the stirps Craniiformis by Zeller and Smith (1964) because of the large sterile base, general nonviolaceous coloration and the cottony texture of the mature gleba.

Scleroderma bovista Fr., *Syst. Mycol.* 3: 48. 1829.

Scleroderma lycoperdoides var. *reticulatum* Coker and Couch, *The Gasteromycetes of the Eastern United States and Canada* 170. 1928.

Fruiting Body: ⅝–1¾″ (1.5–4.5 cm) wide, more or less spherical, becoming flattened on the upper surface in age, attached to the substrate by a thick stalk-like mycelium.

Peridium: firm, dry, smooth when young, soon developing fine cracks and typically divided into small patches 1⁄16–¼″ (2–6 mm) wide, slightly to markedly squamulose in age, straw-yellow to pale orange-yellow when young, becoming orange-yellow to reddish brown with olive-gray tints at maturity, 1⁄32–1⁄16″ (0.8–2 mm) thick, splitting irregularly over the top and sides at maturity.

Base: with dense mycelium forming a stalk-like, soil-filled base ¼–1⅜″ (5–35 mm) long, ½–1⅛″ (1.2–3 cm) thick, straw-yellow to pale orange-yellow, composed of a dense mass of entangled mycelioid cords with trapped sand.

Spore Mass: composed of yellowish peridioles delineated by a thin whitish membrane, with dark blackish brown firm contents at first, powdery and dingy blackish brown in age.

Technical Features: **tramal plates** yellowish; **hyphae** with distinct clamp connections; **spores** 9–16 µm, globose, with a well developed reticulum which is imperfectly developed on some; the reticulum measures up to 3 µm high.

Fruiting: solitary, scattered or in groups on soil in grassy areas, mulch beds, waste areas or woodlands, especially oak-pine; August-December; widely distributed in eastern North America; frequent, but especially common in Florida and the Gulf Coast region.

Edibility: poisonous.

Observations: The more or less spherical, yellow to reddish brown fruiting body, yellow to orange-yellow mycelial base, and a firm dark blackish brown gleba with yellowish peridioles are distinctive features. It is a mycorrhizal species ofen found in dry open areas. *Scleroderma floridanum* Guzmán (1970) is similar but has a smooth peridium which splits in a somewhat star-shaped pattern and spores which are subreticulate.

Notes: All species of *Scleroderma* are poisonous and should be avoided.

Tremella reticulata (Berk.) Farl., *Rhodora* 10: 12. 1908.

Fruiting Body: 1⅛–3⅛″ (3–8 cm) tall, up to 4¾″ (12 cm) wide, composed of finger-like, erect hollow lobes, cristate at first with blunt apices when mature, more or less fused together with a firm, gelatinous, glabrous surface, white at first becoming pale orange-buff to cream color in age.

Hymenium: smooth, covering the entire exposed surface.

Flesh: firm, gelatinous, with fused and reticulate arms when sectioned; odor not distinctive; taste mild.

Technical Features: probasidia oval to pyriform, 11–15 × 8–11 µm, hyaline, becoming cruciate-septate (4-celled); **spores** 7–11 × 5–7 µm, broadly elliptical to broadly ovate, thin-walled, depressed on one side.

Spore Print: white.

Fruiting: growing on debris on the ground or on well rotted stumps and logs, especially oak; in the late summer and fall; distributed in the eastern United States and Canada as far west as Iowa; infrequent.

Edibility: unknown.

Observations: The erect, lobed, fruiting body is very distinctive among the species of jelly fungi. Martin (1944, Plate V) illustrates the fused and reticulate nature of the fruiting body when sectioned as well as the cristate nature of the young, immature lobes. The specimen illustrated here is mature and has blunt, hollow, finger-like lobes and the typical cream color in age.

Calostoma lutescens (Schwein.) Burnap, *Bot. Gaz.* 23: 180–196. 1897.

Mitromyces lutescens Schwein., *Syn. Car.* p. 60. no. 345. 1822.

Fruiting Body: head globose, pale- to lemon-yellow, ½–⅞″ (1.5–2 cm) in diameter; exoperidium weakly gelatinous, breaking up to form a torn ring around the bottom of the globose head; mesoperidium thin, pale yellow, sloughing off with the exoperidium and revealing the dull yellow endoperidium and the raised, red, star-shaped peristome; flesh a white spore sac suspended from and attached to the peristome.

Stalk: 2–3½″ (5–9 cm) long, ⅝–⅞″ (1.5–2 cm) wide, composed of regularly arranged, longitudinal cords which branch and anastomose; leathery when fresh, hard when dry, pale yellow.

Spore Mass: pure white within the suspended sac, attached to the peristome; odor and taste unknown.

Technical Features: exoperidium of interwoven, encrusted hyphae 3–4.2 μm diameter; **mesoperidium** a refractive layer of large interwoven, thick-walled hyphae 8.5–11 μm diameter;

endoperidium of two layers with the outer layer of tightly interwoven, branched hyphae 1.7–3.5 μm diameter and an inner layer of interwoven hyphae 1.7–2.5 μm diameter embedded in a refractive gelatinous matrix; **clamp connections** absent; **spore sac** composed of sparsely branched, interwoven, thick-walled hyphae 3–6.5 μm diameter; **basidia** collapsing at maturity and not observed; **spores** 5.5–8 × 5.5–8 μm, globose, pitted but appearing like reticulate ridges, pale yellow in 3% KOH.

Fruiting: hypogeous or partially hypogeous in soil, emerging at maturity, in deciduous or mixed coniferous-deciduous woods, often in disturbed areas along trails or exposed soil; in the fall and winter months from September to March, but most often in the fall; found in eastern North America from New England south into Mexico; frequent.

Edibility: inedible.

Observations: The yellow spore case which remains at maturity as a persistent torn ring surrounding the head, and the globose pitted spores are a unique combination of characters which distinguish *C. lutescens*, (Miller and Miller 1988, Figs. 31–33; Castro-Mendoza et al. 1983). The other two North American species are *C. cinnabarina* Desv. with a red spore case (Miller 1973, Fig. 369) and *C. ravenelii* (Berk.) Mass. with a thin, gelatinous spore case which becomes granular in age and elliptical spores (10–17 × 6.5–7.5 μm). A fourth taxon is a small spored variety, *C. ravenelii* var. *microsporum* (Atk.) E. Castro-Mendoza and O. K. Miller with spores which measure 5–7 × 3–4 μm but does not differ in any other way. Several strains with intermediate spore sizes have now been recorded in this complex.

Pseudocolus fusiformis (Fischer) Lloyd, *Synopsis of the Known Phalloids*, 53. 1909.

Fruiting Body: ¾–1⅜″ (2–3.5 cm) diameter, at first resembling a small puffball, oval to pear-shaped, grayish brown to pale gray or rarely whitish, typically finely areolate on the upper portion, soon splitting open to form a stalk with tapering arms, a volva and spore mass.

Stalk: ¾–1¾″ (2–4.5 cm) long, shorter or equal in length to the arms, ⅝–1⅛″ (1.5–3 cm) thick, hollow, giving rise to 3–5 reticulate-pitted, arched arms which taper upward and are often united at their apices, whitish at the base, becoming yellow to orange above; arms 1–2¾″ (2.5–7 cm) long.

Volva: grayish brown to pale gray or rarely whitish, typically finely to coarsely areolate on the upper portion with white showing in the cracks.

Spore Mass: olive-green to dark green, borne on the inner side of the arms, slimy, drying nearly black; odor foul-smelling; taste unknown.

Technical Features: spores 4.5–5.5 × 2–2.5 µm, ellipsoid-ovoid, smooth, hyaline.

Fruiting: scattered or in groups on soil in conifer or mixed woods, in wood chips used in gardens or for landscaping; July-September; New York and New England south to North Carolina and spreading; infrequent to occasional.

Edibility: unknown.

Observations: This species was first reported in North America in Pennsylvania in 1915. Collections have been made from several parts of the world including Australia, Japan, Java, the Philippines, Reunion Island, and the United States. *Pseudocolus* can be separated from other stinkhorns by a combination of a stalk shorter or equal in length to the arms which arch and taper to a common point, 3–5 fertile arms and an olive-green gleba on the inner faces of the arms.

Notes: This species is also known as *Colus schellenbergiae* Sumstine, and *Pseudocolus schellenbergiae* (Sumstine) Johnson. It is sometimes called the "Stinky Squid." For additional information see Blanton (1976). For information about *Pseudocolus* in the literature, see Blanton and Burk (1980).

Gyromitra sphaerospora (Pk.) Sacc., *Sylloge Fungorum Omnium Hucusque* 8: 16. 1889.

Cap: 2–4¾″ (5–12 cm) wide, 1⅜–3⅛″ (3.5–8 cm) tall, brain- to saddle-shaped or irregular, wrinkled to convoluted, dry; margin undulating to contorted, strongly incurved when young and typically remaining so at maturity; fertile surface yellow-brown to grayish brown; sterile surface whitish, slightly tomentose.

Stalk: 1⅜–3½″ (3.5–9 cm) long, ¾–2″ (2–5 cm) thick, stout, variable, typically enlarged toward the base and apex, sometimes nearly equal; surface purplish red to rosy pink with whitish areas when young, becoming pinkish brown to pale brownish yellow in age; deeply pitted and ribbed, with ribs markedly extending onto the sterile surface of the cap, smooth on the upper portion, with a fine, white basal tomentum, or smooth overall.

Flesh: thin and brittle; odor and taste unknown.

Technical Features: asci 8-spored, cylindric or subcylindric, 195–220 × 11–16 µm, thin-walled, operculate; **paraphyses** clavate to abruptly enlarged at the apex, 4–12 (–18) µm; **spores** 9–12 µm, globose, smooth, uniseriate.

Fruiting: solitary, scattered or in groups on decaying logs and stumps in hardwoods or mixed woods or on sawdust piles; May-July; eastern Canada, New York, Vermont, New Hampshire, Michigan, Montana; rare.

Edibility: unknown.

Observations: The purplish red to rosy pink stalk and round spores are diagnostic of this species. *Gyromitra californica* (Phillips) Raitviir is a nearly identical poisonous species but larger, has elliptical smooth spores that measure 15–18 × 8–9 μm, and grows on soil in conifer woods in the Rocky Mountains, the Pacific Northwest and Pacific Southwest.

Notes: This species is also known as *Helvella sphaerospora* Pk. and the "Round-spored Gyromitra." For additional information see Pk. (1875) and Weber (1988, 159–161).

Helvella sulcata Fr., *Syst. Mycol.* 2: 15. 1822.

Cap: ⅝–2¾″ (1.5–7 cm) wide, saddle-shaped to trilobate or irregularly convoluted; upper surface brownish black and somewhat shiny when young, becoming dull gray and smooth to somewhat wrinkled in age, dry to slightly slippery; lower surface pale gray, smooth; margin variable, free and straight or curved toward or away from the stalk, often uplifted and wavy or contorted.

Stalk: ¾–2½″ (2–6.5 cm) long, ¼–1″ (6–25 mm) thick, nearly equal overall, pale gray to brownish gray, deeply fluted, branching and anastomosing, occasionally pitted.

Flesh: thin, brittle; odor none; taste unknown.

Technical Features: **asci** 8-spored, cylindric to subcylindric, 90–240 × 12–16 µm, thin-walled; **paraphyses** filiform, becoming clavate at the apex, 4–9 µm wide; **spores** 15–18 × 10–12.5 µm, ellipsoid, with one oil drop, smooth, uniseriate.

Fruiting: solitary, scattered, or in groups on soil, among mosses or on decaying wood in mixed woods; June to October; eastern Canada south to Pennsylvania, west to Michigan, and the Pacific Northwest; infrequent to occasional.

Edibility: unknown.

Observations: The combination of a brownish black to gray, saddle-shaped to convoluted cap and grayish, deeply fluted stalk is distinctive. *Helvella lacunosa* Fr., an edible species, is similar but has an irregularly mitrate to convex cap and a pitted and fluted stalk.

Notes: For more information see Weber (1972, 183–186).

Wynnea americana Thaxter, *Bot. Gaz.* 39: 246. 1905.

Fruiting Body: 1–5½″ (2.5–14 cm) wide, 2⅜–5⅛″ (6–13 cm) high, composed of several to as many as 24 apothecia of variable size but appearing as elongated rabbit ears up to 5½″ (13 cm) high, arising from a tough, irregularly lobed, brown underground sclerotium which measures 1½–2″ (4–5 cm) wide.

Fertile Surface: pinkish orange to pinkish red or brownish orange, smooth.

Sterile Surface: blackish brown to reddish brown, covered with small rounded warts, sometimes wrinkled at maturity.

Flesh: firm, somewhat tough, brown; odor none; taste unknown.

Technical Features: **asci** cylindric, 500–540 µm long, up to 18 µm wide, thin-walled; **paraphyses** filiform, becoming clavate at the apex, septate; **spores** 32–40 × 15–16 µm, elliptical, extremities apiculate, striately marked by several alternately light and dark bands extending the length of the spore, containing oil-drops.

Fruiting: solitary or scattered in moist, organic soils under hardwoods and in disturbed habitats in the southeast referred to as "cove hardwoods"; July-September; New York to Michigan south to Tennessee and North Carolina; infrequent.

Edibility: unknown.

Observations: The distinctive cluster of elongated cups (rabbit ears) combined with the red to orange hymenium are the unique features. *Wynnea sparassoides* Pfister (1979) has a long solid stalk embedded in soil and a rounded brain-like to cauliflower-shaped, yellow-brown head measuring up to 3⅛″ (8 cm) in diameter.

Notes: For additional information see Seaver (1928).

Glossary

Works Cited

Index

Glossary

acrid: burning or peppery tasting

acuminate: gradually narrowed to a point

adnate: broadly attached to the stalk

allantoid: curved and sausage-shaped

amyloid: staining blue, blue-gray to blackish violet in Melzer's reagent

anastomosing: connecting crosswise to form a vein-like network

annular ring: a distinct ring

annular zone: an indistinct ring or zone of fibrils

annulus: a ring of tissue left on the stalk by the torn partial veil

apex: uppermost portion of the stalk or the very top part

apical pore: a small opening located at the spore apex

apices: tips, outermost portions

apiculate: having a short projection

areolate: marked out into small areas by cracks or crevices

aromatic: having an agreeable pungent aroma

asci: the sexual reproductive cells in which ascospores are produced

attenuate: gradually narrowed

azonate: lacking zones

basal mycelium: an entangled mass of hyphae located at the base of the stalk

basidia: the sexual cells of Basidiomycetes which give rise to basidiospores

bilateral trama: with hyphae diverging like an inverted V on either side from the center of the gill

binding hyphae: a type of hyphae with short branches found in some polypores which provides strength to the tissue

boletinoid: radially arranged and elongated pores

bulbous: enlarged at the base of the stalk

buttons: undeveloped immature mushrooms

campanulate: bell-shaped

canescence: a pale hoary down

cap: the part of a mushroom which is supported by the stalk and which supports the gills, tubes, spines, etc.

capillitium: sterile, thread-like, often branched, thick-walled hyphae in the gleba (spore mass) of puffballs, earthstars, etc.

capitate: having a small knob at the apex of a cystidium

caulocystidia: sterile cells or hairs located on the surface of the stalk

cheilocystidia: sterile cells located on the gill edge

clamp connections: small semicircular branches at the septum on hyphae of many, but not all, Basidiomycetes

clamps: *see* clamp connections

clavate: club-shaped

clavate-mucronate: club-shaped and tipped with an abrupt, short, sharp point

close: the spacing of gills halfway between crowded and subdistant

conchate: shell-shaped

conifer: a cone bearing tree such as pine or spruce

coremia: minute asexual structures with a white stalk and a black-brown head

coriaceous: having a leathery texture

cortina: a silky, web-like partial veil

crenulate: very finely scalloped

cruciate: having the general appearance of a cross

cystidia: large sterile cells which may project from the cap, gills, or stalk

decurrent: descending down onto the stalk

dehiscence: opening by a slit or pore or by tearing

deliquescent: liquefying at maturity

depressed: having the disc lower than the margin of the cap or tubes, or sunken around the stalk

dextrinoid: dark red to red-brown color when placed in Melzer's solution

dimitic: having generative and either skeletal or binding hyphae

disc: the central portion of the surface of the cap

divergent hyphae: *see* bilateral trama

duff: cast off needles and debris which accumulates on the forest floor

eccentric: attached off center

elliptical: having the shape of an ellipse, rounded on both ends and with curved sides

emarginate: notched near the stalk

end cells: short, thick-walled, terminal hyphae

endoperidium: the innermost enveloping layer

farinaceous: having an odor of fresh meal

fascicles: small bundles

fertile surface: the area on which reproductive structures are found

$FeSO_4$: iron sulfate used for a specific stain on fresh flesh

fibers: thin thread-like filaments

fibrillose-scaly: having scales composed of fibrils that are more or less flattened

fibrils: minute thread-like fibers

filiform: slender like a thread

fimbriate: finely torn or fringed

flabelliform: fan-shaped

floccose: having tufts of soft, cottony material

floccose-membranous: having a combination of cottony and membrane-like features

fluted: having sharp-cornered ridges extending down the stalk

fruiting body: the entire portion of a mushroom developed for producing spores

furrows: narrow grooves

fusiform-ellipsoid: elliptical but somewhat spindle-shaped

fusoid-ventricose: swollen in the middle and somewhat spindle-shaped with the apex rounded or acute

gelatinized: having a jelly-like consistency

generative hyphae: simple, thin-walled hyphae which differentiate to form tissues and cells in the fruiting body

germ pore: a thin portion of the spore wall through which the germ tube grows during germination

glabrous: smooth, bald, lacking scales, fibrils, etc.

glandular dots: sticky drops or glands composed of caulocystidia, located on the stalk of *Suillus,* for example

gleba: a spore mass contained within a peridium in the Gasteromycetes such as Lycoperdales, Tulostomatales, etc.

globose: round or spherical

gluten: a sticky glue-like pectinous material

hemispherical: shaped like one-half of a sphere

hirsute: covered with a dense layer of long, stiff hairs

homogeneous: having uniform composition

hyaline: transparent, colorless

hygrophanous: watery and typically changing color when fading

hyphae: microscopic thread-like filaments

hyphal pegs: small hyphal bundles projecting out from the spore-bearing surface

hypogeous: arising beneath the surface of the earth

imbricate: overlapping like shingles

inamyloid: a yellow to hyaline reaction in Melzer's solution

incurved: bent inward

indehiscent: not opening along specific lines or patterns

inrolled: rolled up and inward

intersterile: unable to mate and reproduce

intervenose: having conspicuous ridge-like veins between gills or ridges

interwoven: see mixocutis

ixolattice: a pileipellis of branching, ascending, entangled hyphae in a gelatinous matrix

ixomixocutis: a viscid pileipellis with interwoven filamentous hyphae

ixotrichodermium: a gelatinized pileipellis with the distal portion of the filiform elements of unequal length and arranged perpendicularly to the surface

KOH: potassium hydroxide, usually made up in a 3% or 10% concentration in water

lacerated: appearing torn or shredded

lactifers: hyphal elements, usually oil filled, containing a milklike fluid

lageniform: flask-shaped

lamellar trama: the supporting sterile, internal tissue of the gill

lamellulae: short gills which do not reach the stalk

latex: a juice which may be milky or watery

lobate: composed of rounded lobes

longitudinally striate: having minute furrows or lines which extend up and down the stalk

lubricous: smooth and slippery

margin: edge of the cap or gills

mediostratum: the central portion of gills or tubes

Melzer's reagent: an iodine based reagent used to test for the amyloid (blue) reaction of fungal spores and cell walls

Melzer's solution: contains 20 cc H_2O, 1.5 g potassium iodide, 0.5 g iodine and 20 g chloral hydrate

mesoperidium: the middle peridial layer of puffballs

metuloids: modified cystidia often encrusted with lime

mitrate: mitre-shaped, resembling a bishop's cap

mixocutis: a pileipellis composed of interwoven hyphae

monomitic: composed of only generative hyphae

mucronate: tipped with an abrupt, short, sharp point

multipileate: having more than one cap

mycelium: an entangled mass of hyphae

mycorrhizal: a mutualistic relationship between the mycelium of a fungus and the tiny, short roots of green plants

nitrous: having a specific chemical odor resembling gun powder

obovate: with the broader portion toward the apex

obtuse: rounded or blunt

ochraceous: having shades of orange-yellow

operculate: having a small lid

palisade: an arrangement of elongated, perpendicular cells

paraphyses: sterile supporting filaments located between asci in the Ascomycetes

partial veil: the inner veil covering the gills or pores

PDAB: para-dimethylamino-benzaldehyde, used to detect a pink staining reaction for some species of *Tricholoma*

pedicel: a thin stalk

pellicle: a skin-like outer covering of the cap which is typically gelatinous and is often separable from the underlying tissues

peridiole: small chambers in the gleba delineated by thin, white to yellow membranes, see description of *Scleroderma bovista*

peridium: the enveloping coat or outer skin of fruiting bodies such as puffballs

peristome: the circular area surrounding the opening to the gleba of some Gasteromycetes

pileipellis: the layers of the outer surface of the cap

pileocystidia: cystidia located on the surface of the cap

pileotrama: the supporting tissue of the cap located between the pileipellis and the spore-bearing surface

plage: a smooth area near the point of attachment on a spore

plane: having a flat surface or flat at maturity

pleurocystidia: cystidia located on the face of a gill or tube

plicate: folded and resembling a fan

pore: a small opening which may be round, angular, or elongated

pore surface: the layer formed by the outermost end or mouths of the tubes

primordial: original, immature

probasidium: a primary cell found in some fungi, such as jelly fungi, which gives rise to a mature basidium

pruinose: covered with a white, powdery substance

pseudocolumella: a stalk-like sterile structure found in the center of the spore mass

pubescent: covered with a layer of short, soft, downy hair

pungent: sharp and piercing odor

pyramidal warts: pyramid-shaped warts

pyriform: pear-shaped

radially striate: having minute furrows or lines spreading from a common center

repent: prostrate

reticulate-pitted: with lines and ridges forming a network of shallow depressions

reticulum: a net-like arrangement or network

rhizomorph: a cord or strand composed of mycelium which penetrates the substrate

rimose: having tiny cracks or crevices

rugose: coarsely wrinkled

scabrous-dotted: adorned with scattered short projections that appear as spots

scales: projections or tearing of the cap or stalk surface forming small, flattened, or erect decorations

scalloped: having a series of shell-shaped curves

sclerotium: a hardened mass of hyphae which serves as a resting body that allows a fungus to survive adverse conditions

septate: hyphal cells divided by crosswalls

serrate: toothed, like the edge of a saw blade

setiform: bristle-shaped

siderophilous granulations: dark purple to purplish black particles observed in the basidia of some fungi which have been stained with acetocarmine

silky-fibrillose: having tiny silky fibrils

sinuate: having a concave depression near the stalk

skeletal hyphae: a type of thick-walled hyphae found in the Aphyllophorales (polypores, hydnums, etc.)

spermatic: resembling the odor of semen

sphaerocysts: a cluster of more or less globose cells

squamulose: covered with minute scales

stalk: the stem of a fungus which supports the cap

stipe-columella: a sterile stalk-like structure inside the fruiting body of some Gasteromycetes

striate: having minute radiating furrows or lines

subdistant: gill spacing halfway between close and distant

subfusiform: more or less spindle-shaped

subgleba: the sterile portion beneath the gleba

sublamellate: resembling gills

substrate: food source on which the fungus is growing such as decaying wood, pine cones or leaves

subventricose: somewhat swollen in the middle

sulcate: grooved

superior: located on the upper portion of the stalk

suprahilar depression: see plage

terete: rod-like, as a broom handle

tiers: rows

tomentose: densely matted and wooly

translucent-striate: the appearance of striations caused by gill edges visible through unusually thin cap tissue

trichodermium: a pileipellis with the distal portion of the filiform elements of equal length and arranged perpendicularly to the surface

trilobate: having three lobes

truncate: having an end appearing cut off

tubercles: wart-like or knob-like projections

tuberculate: with small projecting warts or knobs

tubes: tiny hollow cylinders in which spores of boletes or polypores are produced

turf: superficial mycelial growth

umbo: a conical to convex elevation on the center of the cap

umbonate: having an umbo

uniseriate: occurring in a single row

universal veil: a tissue which envelops the young fruiting body of some fungi

vegetative hyphae: actively growing and feeding hyphae

ventricose: swollen in the middle and tapering in either direction

ventricose-rostrate: swollen in the middle with a narrow tip

verrucose: covered with small warts

vesiculose: bladder-like or somewhat spherical

villose: having long, soft hairs

virgate: streaked with fibrils

viscid: sticky to the touch

volva: the remains of the universal veil located at or surrounding the base of the stalk

zonations: concentric bands, often of different colors

Works Cited

Askew, B., and O. K. Miller. 1977. New Evidence of Close Relationships Between *Radiigera* and *Geastrum*. *Can. J. Bot.* 55: 2693–2700.

Bigelow, H. E. 1982. *North American Species of Clitocybe. Part 1.* J. Cramer, Berlin. 280 pp.

———. 1985. *North American Species of Clitocybe. Part 2.* J. Cramer, Berlin. 240 pp.

Blanton, R. L. 1976. *Pseudocolus fusiformis*, New to North Carolina. *Mycologia* 68: 1235–1239.

Blanton, R. L., and W. R. Burk. 1980. Notes on *Pseudocolus fusiformis*. *Mycotaxon* 12: 225–234.

Breitenbach, J., and F. Kränzlin. 1991. *Fungi of Switzerland. Vol. 3.* Sticher Printing, Lucern, Switzerland. 361 pp.

Canfield, E. R. 1981. The Wood-Decay Capability of *Albatrellus dispansus*. *Mycologia* 73: 399–406.

Canfield, E. R., and R. L. Gilbertson. 1971. Notes on the Genus *Albatrellus* in Arizona. *Mycologia* 63: 964–971.

Castellano, M. A., J. M. Trappe, Z. Maser, and C. Maser. 1989. *Key to Spores of the Genera of Hypogeous Fungi of North Temperate Forests (with Special Reference to Animal Mycophagy).* Mad River Press, Eureka, Calif. 186 pp.

Castro-Mendoza, E., O. K. Miller Jr., and D. A Stetler. 1983. Basidiospore Wall Ultrastructure and Tissue System Morphology in the Genus *Calostoma* in North America. *Mycologia* 75: 36–45.

Cetto, B. 1979. *Der Pilzführer, Band 3*. B3V, Verlags Gesellschaft, München. 275 pp.

Coker, W. C., and J. N. Couch. 1928. *Gasteromycetes of the Eastern United States and Canada.* Univ. of North Carolina Press, Chapel Hill. 201 pp.

Donk, M. A. 1962. *The Generic Names Proposed for Agaricaceae.* Beih. Nova Hedw., Heft 5, Cramer, Weinheim, Germany. 320 pp.

Døssing, L. 1979. *Ramicola* Vel., 339–340. In *Nordic Macromycetes.* Vol. 2., L Hansen and H. Knudsen. Nordsvamp, Copenhagen. 474 pp.

Gilbertson, R. L., and L. Ryvarden. 1986. *North American Polypores. Vol. 1.* Fungiflora, Oslo. 433 pp.

———1987. *North American Polypores. Vol. 2.* Fungiflora, Oslo. 434–885 pp.

Gillman, L. S., and O. K. Miller. 1977. A Study of the Boreal, Alpine, and Arctic Species of *Melanoleuca. Mycologia* 69: 927–951.

Grund, D. W., and D. E. Stuntz. 1981. Nova Scotian Inocybes 6. *Mycologia* 73: 655–674.

Guzmán, G. 1970. Monografia del Genero *Scleroderma. Darwiniana* 16: 233–407.

Hall, D., and D. E. Stuntz. 1971. Pileate Hydnaceae of the Puget Sound Area. 1. White-Spored Genera: *Auriscalpium, Hericium, Dentinum* and *Phellodon. Mycologia* 48:1099–1128.

Hansen, L., and H. Knudsen. 1992. *Nordic Macromycetes Vol. 2.* Nordsvamp, Copenhagen. 474 pp.

Harmaja, H. 1979. Studies in the Genus *Cystoderma. Karstenia* 19: 25–29.

Harrison, K. A. 1961. *Stipitate Hydnums of Nova Scotia.* Canadian Dept. of Agriculture, Ottawa. 60 pp.

———1964. New or Little Known North American Stipitate Hydnums. *Can. J. Bot.* 42: 1205–1233.

Hawksworth, D. L., B. C. Sutton, and G. C. Ainsworth. 1983. *Dictionary of the Fungi. 7th Ed.* Commonwealth Mycological Institute, Kew, England. 448 pp.

Hesler, L. R. 1969. *North American Species of Gymnopilus.* Hafner, Darien, Conn. 117 pp.

Hesler, L. R., and A. H. Smith. 1963. *North American Species of Hygrophorus.* Univ. of Tennessee Press, Knoxville. 416 pp.

———1979. *North American Species of Lactarius.* Univ. of Michigan Press, Ann Arbor. 841 pp.

Jenkins, D. T. 1977. A Taxonomic and Nomenclatural Study of the Genus *Amanita* section *Amanita* for North America. *Bibliotheca Mycologica* 57:1–127.

———1986. *Amanita of North America.* Mad River Press, Eureka, Calif. 198 pp.

Jülich, W. 1981. Higher Taxa of Basidiomycetes. *Bibliotheca Mycologica* 85. J. Cramer, Vaduz, Germany. 485 pp.

Kuyper, T. 1986. *A Revision of the Genus Inocybe in Europe. I. Subgenus Inosperma and the Smooth-spored Species of Subgenus Inocybe.* Rijksherbarium, Leiden. 247 pp.

Lange, J. E. 1938. *Flora Agaricina Danica 4*. Recato, Copenhagen. 119 pp.

Largent, D. L. 1985. *The Agaricales (Gilled Fungi) of California 5. Hygrophoraceae*. Mad River Press, Eureka, Calif. 208 pp.

Lincoff, G. H. 1981. *The Audubon Society Field Guide to North American Mushrooms*. Knopf, New York. 498 pp.

Marr, C. D., and D. E. Stuntz. 1973. *Ramaria* of Western Washington. *Bibliotheca Mycologica* 38: 1–232.

Martin, G. W. 1944. The Tremellales of the North Central United States and Adjacent Canada. *Univ. of Iowa Studies in Nat. Hist.* 18(3): 1–88.

Miller, O. K. 1965a. Three New Species of Lignicolous Agarics in the Tricholomataceae. *Mycologia* 57: 933–945.

——— 1965b. Snowbank Mushrooms in the Three Sister Wilderness Area. *Mazama* 47: No. 13. 38–41.

——— 1969. A New Species of *Pleurotus* with a Coremioid Imperfect Stage. *Mycologia* 61: 887–893.

——— 1971. The Genus *Gomphidius* with a Revised Description of the Gomphidiaceae and a Key to the Genera. *Mycologia* 63: 1129–1163.

——— 1973. *Mushrooms of North America*. E. P. Dutton, New York. 360 pp.

——— 1993."Observations on the Genus *Cystoderma* in Alaska." In *Arctic and Alpine Mycology* 3, edited by O. Petrini and G. A. Laursen, 161–169, *Bibliotheca Mycologia* 150: 1–269

Miller, O. K., and H. H. Miller. 1980. *Mushrooms in Color*. E. P. Dutton, New York. 286 pp.

——— 1988. *Gasteromycetes, Morphological and Development Features with Keys to the Orders, Families and Genera*. Mad River Press, Eureka, Calif. 157 pp.

Miller, O. K., and L. Stewart. 1971. The Genus *Lentinellus*. *Mycologia* 63: 333–369.

Miller, O. K., E. Trueblood, and D. Jenkins. 1990. Three New Species of *Amanita* from Southwestern Idaho and Southeastern Oregon. *Mycologia* 82: 120–128.

Morse, E. E. 1930. A New Chanterelle in California. *Mycologia* 22: 219–220.

Moser, M. 1983. *Keys to Agarics and Boleti*. Gustav Fischer Verlag, Stuttgart. 535 pp.

Mueller, G. M. 1992. Systematics of *Laccaria* (Agaricales) in the Continental United States and Canada, with Discussions on Extralimital Taxa and Descriptions of Extant Types. *Fieldiana* Pub. 1435 n. s. no. 30. 158 pp.

Murrill, W. A. 1916. Omphalina. *North American Flora* 9: 344–352.

Ovrebo, C. L. 1989. *Tricholoma,* Subgenus *Tricholoma,* Section *Albidogrisea*: North American Species Found Principally in the Great Lakes Region. *Can. J. Bot.* 67: 3134–3152.

Overholts, L. O. 1953. *The Polyporaceae of the United States, Alaska and Canada*. Univ. of Michigan. Press, Ann Arbor. 466 pp.

Peck, C. H. 1873. *Agaricus (Tricholoma) decorosus*. Bull. Buffalo Soc. Nat. Sci. 1: 42–43.

——— 1875. Ann. Rept. N.Y. Sta. Museum 27: 106.

Pfister, D. 1979. A Monograph of the Genus *Wynnea* (Pezizales, Sarcoscyphaceae). *Mycologia* 71: 144–159.

Phillips, R. 1991. *Mushrooms of North America.* Little, Brown, Boston. 319 pp.

Pollack, F. and O. K. Miller. 1976. *Antromycopsis broussonetiae* Found To Be the Name of the Imperfect State of *Pleurotus cystidiosus*. *Mem. N.Y. Bot. Gdn.* 28: 174–178.

Pomerleau, R. 1964. An Addition to the Genus *Fuscoboletinus*. *Mycologia* 56: 708–711.

Redhead, S. A. 1987. Notes on the Genus *Xeromphalina* (Agaricales, Xerulaceae) in Canada: Biogeography, Nomenclature, Taxonomy. *Can. J. Bot.* 66: 479–507.

Redhead, S. A., and J. H. Ginns. 1985. A Reappraisal of Agaric Genera Associated with Brown Rots of Wood. *Trans. Mycol. Soc. Japan* 26: 349–381.

Seaver, F. J. 1928. *The North American Cup-fungi. (Operculates)*. Hafner, New York. 284 pp.

Singer, R. 1943. Type Studies on Basidiomycetes 2: Tricholomataceae. *Mycologia* 35: 152.

Smith, A. H. 1937. Studies in the Genus *Mycena*. *Mycologia* 29: 338–354.

——— 1944. New and Interesting Cortinarii from North America. *Lloydia* 7(3): 163–235.

——— 1947. *The North American Species of Mycena*. Univ. of Michigan Press, Ann Arbor. 521 pp.

——— 1949. *Mushrooms and Their Natural Habitats*. Sawyer's , Portland, Ore. 626 pp.

Smith, A. H., and L. R. Hesler. 1968. *The North American Species of Pholiota*. Hafner, New York. 402 pp.

Smith, A. H., and E. E. Morse. 1947. The Genus *Cantharellus* in the Western United States. *Mycologia* 39: 497–534.

Smith, A. H., H. V. Smith, and N. S. Weber. 1979. *How to Know the Gilled Mushrooms*. William. C. Brown, Dubuque, Iowa. 334 pp.

Smith, A. H., and H. D. Thiers. 1964. *A Contribution Toward a Monograph of North American Species of Suillus*. Ann Arbor. 116 pp.

Smith, A. H., and H. D. Thiers. 1971. *The Boletes of Michigan*. Univ. of Michigan. Press, Ann Arbor. 428 pp.

Smith, A. H., H. D. Thiers, and O. K. Miller. 1965. The Species of *Suillus* and *Fuscoboletinus* of the Priest River Experimental Forest and Vicinity, Priest River, Idaho. *Lloydia* 28: 120–138.

Smith, H. V., and A. H. Smith. 1973. *How to Know the Non-Gilled Fleshy Fungi*. William. C. Brown, Dubuque, Iowa. 402 pp.

Stamets, P. 1993. *Growing Gourmet and Medicinal Mushrooms*. Ten Speed Press, Berkeley, Calif. 552 pp.

Tylutki, E. E. 1987. *Mushrooms of Idaho and the Pacific Northwest*. Vol. 2. *Non-Gilled Hymenomycetes*. Univ. of Idaho Press, Moscow, Idaho. 232 pp.

Velenovský, J. 1920. *Ceské Houby*, Muskem Edu. Leschengr v. Praze, Czech. 950 pp.

Vilgalys, R., A. Smith, B. L. Sun, and O. K. Miller. 1992. Intersterility Groups in the *Pleurotus ostreatus* Complex from the Continental United States and Adjacent Canada. *Can. J. Bot.* 71: 113–128.

Volk, T. J., H. H. Burdsall, and K. Reynolds. 1994. Checklist and Host Index of Wood-Inhabiting Fungi of Alaska. *Mycotaxon* 52: 1–46.

Weber, N. S. 1972. The Genus *Helvella* in Michigan. *Mich. Bot.* 11: 183–186.

Weber, N. S. 1988. *A Morel Hunter's Companion*. Two Peninsula Press, Lansing, Mich. 209 pp.

Zeller, S. M. 1944. Representatives of the Mesophelliaceae in North America. *Mycologia* 36: 627–637.

Zeller, S. M., and A. H. Smith. 1964. The Genus *Calvatia* in North America. *Lloydia* 27: 148–186.

Index to Genera and Species

Alan E. Bessette is a mycologist and professor of biology at Utica College of Syracuse University. He has published numerous professional papers in the field of mycology and has authored eight books including: *Edible and Poisonous Mushrooms of New York*, *Mushrooms of the Adirondacks*, *Mushrooms: A Quick Reference Guide to Mushrooms of North America* (coauthored with Walter J. Sundberg) and *Edible Wild Mushrooms of North America* (coauthored with David W. Fischer). Professor Bessette has presented numerous mycological programs, is the scientific advisor to the Mid-York Mycological Society, and serves as a consultant for the New York State Poison Control Center. He has been the principal mycologist at national and regional forays and was the recipient of the 1987 Northeast Mycological Foray Service Award and the 1992 North American Mycological Association Award for Contributions to Amateur Mycology.

Orson K. Miller is professor of botany and curator of fungi in the Department of Biology, Virginia Polytechnic Institute and State University. He is also visiting professor of botany at the Flathead Lake Biological Station of the University of Montana. Over the past 36 years he has collected and studied fungi in Europe, Asia, Australia, and Africa, as well as extensively in North America. He has published over 120 professional papers and book chapters. His five books include: *Mushrooms of North America*; *Index of the Fungi*; *Mushrooms in Color* and *Gasteromycetes*, coauthored with his wife Hope; and *Texas Mushrooms*, coauthored with Susan and Van Metzler. He has been an officer in the Mycological Society of America and was awarded the Weston Award for Excellence in Teaching by the society. He has led many forays for the society, national and international forays for North American Mycological Society (NAMA), and was the 1981 recipient of the NAMA Award for Contributions to Amateur Mycology.

Arleen R. Bessette is an amateur mycologist and botanical photographer, as well as a professional psychologist, who has been collecting and studying wild mushrooms for several years. She created and contributed over fifty original recipes for the book *Edible Wild Mushrooms of North America* and is the author of *Taming the Wild Mushroom: A Culinary Guide to Market Foraging*. Arleen has won several national awards for her photography and has taught introductory courses in mycology for the North American Mycological Association. Together, the Besssettes host regional forays from the Adirondacks to Cape Cod, offering participants the opportunity to collect and study mushrooms, including wild edibles.

Hope H. Miller is the other half of the team of Miller and Miller. She is the author of *Hope's Mushroom Cookbook* which contains 320 recipes, which is the culmination of over 35 years of work in kitchens around the world. She has coauthored two books with her husband, and has received the Lifetime Achievement Award for Contributions to Amateur Mycology from the Texas Mycological Society in 1988. Hope teaches each year for the Open University in Blacksburg on "How to Know Your Mushrooms," and "Cooking with Mushrooms," and lectures to many local organizations, including garden clubs. Hope is frequently found giving cooking demonstrations to mushroom clubs around the country and to mycology classes at Virginia Tech. She has appeared on many TV and radio shows sharing her knowledge, and has served as field assistant for her husband on his many mycological forays and scientific field trips througout the world.